Learning about Risk

Learning about Risk

**Consumer and Worker Responses
to Hazard Information**

**W. KIP VISCUSI and
WESLEY A. MAGAT**

WITH
JOEL HUBER, CHARLES O'CONNOR,
JAMES R. BETTMAN, JOHN W. PAYNE,
AND RICHARD STAELIN

Harvard University Press

Cambridge, Massachusetts
and London, England   1987

Many of the designations used by manufacturers and sellers to distinguish their products are claimed as trademarks. Where those designations appear in this book and Harvard University Press was aware of the trademark claim, the designations have been printed with initial capital letters.

This book is printed on acid-free paper, and its binding materials have been chosen for strength and durability.

Library of Congress Cataloging-in-Publication Data

Viscusi, W. Kip.
   Learning about risk.

   Bibliography: p.
   Includes index.
   1. Consumer protection—United States.
2. Communication in consumer education—United States.
3. Communication in industrial safety—United States.
4. Chemicals—United States—Safety measures.
5. Hazardous substances—United States. I. Magat, Wesley A.
II. Title.
HC110.C63V57 1987   381'.33'0973   86-18448
ISBN 0-674-51915-9 (alk. paper)

PREFACE

Whereas the 1970s marked the emergence of direct government regulation of risks as a major component of government policy, the 1980s have witnessed increased utilization of information programs that would augment existing market forces. The Occupational Safety and Health Administration's most ambitious policy effort since the agency's inception has been to implement hazard communication systems, principally through labeling hazardous chemicals in the workplace. Several states have launched "right-to-know" efforts, and the federal government is attempting to discourage cigarette smoking by rotating hazard warnings on the sides of cigarette packages.

Despite the proliferation of risk information programs in general, and of labeling efforts in particular, relatively little is known about their effects. The literature on advertising and marketing addresses general consumer labeling issues but not the specific concerns raised by risk warnings. The professional chemical labeling literature summarizes existing labeling programs and regulatory requirements but does not consider the ultimate impact of labels. Until now there has been no firm empirical evidence on the impact of hazard warnings other than what can be extrapolated from case studies and from our general knowledge of human behavior.

The research results reported in this book rely on new field studies of consumer and worker behavior and are aimed at two broad audiences. Because these surveys use economics models and test the types of hypotheses most pertinent to economic behavior, they are of interest to economists and to researchers in marketing, decision sciences, and psychology who deal with related issues. The experimental design, statistical tests, and methodology accord with the standards of social science scholarship. Wherever possible, we have attempted to note the relationship of our work to that in other disciplines, and we have devoted an entire chapter to the state of the art regarding risk information in the fields of psychology and marketing.

Our second audience consists of labeling practitioners in both government and industry. We address issues such as the design of an effective labeling system and analyze the diverse economic effects that labels will have. This group of readers will learn much from the general discussion even without focusing on the specifically methodological sections.

Chapter 1 sets the context for the study within the broader literature on decisions under uncertainty. Chapter 2 reviews the implications of the marketing and psychology literature and discusses the rationale for one of the label designs tested in Chapter 4. Researchers in different disciplines can easily gear their reading of the sections in this chapter to their particular interests and backgrounds.

Chapter 3 provides an overview of the design of the consumer information study and the labeling treatments used. This chapter is an essential prerequisite for Chapter 4 and provides useful background information on the sample analyzed in Chapter 5.

Chapter 4 surveys the effects of labels on consumer precautions and the costs to consumers of taking these precautions. Chapter 5 examines how changes in the riskiness of the product affect the price consumers are willing to pay for it. Each of these chapters provides both formal statistical analysis and a summary of the overall average effects for the sample.

Chapter 6 discusses a field study of the effect of hazard warnings in the workplace, ranging from altered risk perceptions to demands for higher wages. Because this chapter also relies on a carefully designed experiment, it is possible to infer the main results from the average effects of the chemical labels on the sample. Chapter 7 summarizes many of the principal themes of the book.

The diversity of disciplinary issues involved in the consumer and worker studies is reflected in the range of expertise of the researchers. The 1982 chemical worker survey by Viscusi and O'Connor (Chapter 6) linked the economic analysis of job hazards reported in Viscusi's earlier work with the introduction of risk information through chemical labels. The format of these labels and the formulation of risk questions that were meaningful to workers owed much to the expertise of Charles O'Connor, then director of chemical labeling at Stauffer Chemical Company. The 1984 consumer labeling study (Chapters 3–5) likewise drew on experts from other disciplines besides economics. Joel Huber (marketing) was closely involved in all phases of the study, ranging from survey design to the intricacies of conjoint analysis. John Payne (psychology), James Bettman (marketing), and Richard Staelin (marketing) wrote a synthesis of what was known in these fields, prepared a test label design, and offered detailed comments on the survey.

The 1984 consumer survey was prepared under Environmental Protection Agency (EPA) Cooperative Agreement no. CR-811057-01-0 with the Center for the Study of Business Regulation in the Fuqua School of Business at Duke University. Viscusi's work was also supported by the Center for the Study of the Economy and the State at the University of Chicago. Because field work of this type involved substantial expense, the continued interest and support of the EPA Office of Policy Analysis were essential. The EPA agreement gave the researchers substantial leeway in formulating their research project, but EPA officials worked closely with them to maintain its policy relevance. This interaction was particularly useful because the EPA is considering revisions in its pesticide labeling regulations and is beginning chemical labeling under the Toxic Substances Control Act.

We would like to thank James Hibbs, of the EPA Office of Policy Analysis, for making smooth the path of the last project he oversaw before his retirement. We also received stimulating comments from Alan Carlin, Ann Fisher, Ralph Luken, Albert McGartland, and Albert Nichols at various stages of the research. Arnold Aspelin, Gary Ballard, and Peter Kuch, of the EPA Office of Pesticide Programs, took an active interest in the project. Sammy Ng and Michael Shapiro, of the EPA Office of Toxic Substances, continually provided thoughtful advice and comment. None of the material in this book represents an official position of the EPA.

Several other individuals made important contributions to this book. Pamela Dressler worked closely with our marketing research firm and was responsible for all the computer programming for the consumer study, and William Evans provided computer programming support for the worker labeling study. We would also like to thank Robert Mitchell for helpful comments on the questionnaire and Gregory M. Duncan for econometric advice. Dr. Shirley Osterhout of the Duke University Poison Control Center was particularly helpful in providing information on the physical effects of exposure to the different product hazards examined.

Turning our research effort into print would not have been possible without the efforts of Corliss Andrews and Terri Copeland, whose mastery of the personal computer accelerated the pace of our successive revisions. Michael Aronson gave encouragement and helpful advice. Howard Kunreuther provided particularly helpful suggestions regarding both the organization of the book and the intepretation of our empirical results.

Abbreviated versions of Chapters 2, 4, and 6 have been published respectively as "Cognitive Considerations in Designing Effective Labels for Presenting Risk Information," *Journal of Public Policy and Marketing,* 5 (1986), 1–28; "Informational Regulation of Consumer Health Risks," *Rand Journal of Economics,* 17 (Autumn 1986), 351–355; and "Adaptive Responses to Chemical Labeling: Are Workers Bayesian Decisionmakers?" *American Economic Review,* 74 (1984), 942–956. We thank these journals for permission to draw upon that material.

W. Kip Viscusi
Wesley A. Magat

CONTENTS

FIGURES

TABLES

Learning about Risk

1 | Information Processing and Individual Decisions

W. KIP VISCUSI and
WESLEY A. MAGAT

For decisions made under conditions of uncertainty, the information individuals have about prospective outcomes and how they use it are central to economic performance. The importance of these factors has led to the establishment of markets and activities for the transfer of information, ranging from advertising to high-tech "information technologies." Inadequacies in individual information have also prompted a wide range of risk regulation efforts for potentially hazardous jobs and products. Although these policies usually take the form of a ban or other technological means of altering the risk, if the chief inadequacy is lack of information, then in principle it should be possible to alleviate the source of the market failure directly with an informational policy.

Hazard warning labels have become a prevalent form of information about risks. Most people take *labeling* to mean a very narrow type of information transfer through a label attached to the outside of a container of, say, a pesticide. In practice, however, labeling includes the leaflets and brochures provided with products, warning signs in the workplace, and point-of-purchase displays. In the following chapters we consider all tangible means of information transfer as labeling. Formal training efforts, such as safety training programs, are excluded, although the lessons drawn here from our studies of labeling information will have some applicability to other informational efforts such as training programs. We view our labeling analysis in broad terms as a case study of the transfer of risk information.

From an economic standpoint, the potential role of labels and other forms of information transfer is identical: all of them influence perceptions of risk and, ultimately, individual behavior. In the design of labels or other forms of information provision, we are concerned with the following set of issues. First, how much information do individuals possess about the risks associated with the decisions they make? Second, can we improve decisionmaking under conditions of uncertainty by providing additional information? To answer this second question we must learn more about how individuals process new information, how this information affects behavior, and how we can devise effective means for transferring information.

Although the literature on the effects of advertising and on the psychology of risk perception is quite extensive, few researchers have examined the most fundamental questions concerning the processing of information and its effect on economic behavior. To remedy these deficiencies, we undertook two field surveys exploring the nature of individual decisionmaking under uncertainty. In the 1984 survey, discussed in Chapters 3, 4, and 5, we interviewed 368 consumers to determine how varying kinds and amounts of information affect precautionary behavior, how design of the information transfer system affects the information individuals acquire, and how this knowledge can be applied to a wide range of related uses. In the 1982 survey, discussed in Chapter 6, we interviewed 355 chemical workers to learn how hazard warnings affect risk perceptions and subsequent worker behavior, such as quit rates and the wage level workers require to remain on the job.

Throughout both studies we tested a paradigm based on an economic approach to decisionmaking under uncertainty. In particular, do individuals maximize expected utility and act as Bayesian decisionmakers when they acquire information, revise their probabilities, and act upon this new information? (For a lucid exposition, see Raiffa 1968.) More specifically, the manner in which individuals use the information they acquire to update their probabilistic beliefs must satisfy Bayes' Theorem, which is an algebraic formulation from probability theory. Consider the following example of this approach. Suppose there are two kinds of jobs in the economy—high-risk jobs, H, and low-risk jobs, L. The worker receives a single piece of information—that the previous employee suffered a fatal job injury. Given this information, the assessed probability, P, that the job is a high-risk job is:

$$P(H|\text{injury}) = \frac{P(\text{injury}|H)P(H)}{P(\text{injury}|H)P(H) + P(\text{injury}|L)P(L)}.$$

The assessed probability that the job is a type H job after the worker learns the fate of the previous employee is given by a quite specific relationship. This "posterior" probability (that is, probability after information is acquired) equals the overall relative chance that an injury could have occurred if the job was in fact hazardous, weighted by the worker's assessed probability that the job is high risk, as compared with the overall risk of injury. When an injury is just as likely with a high-risk as with a low-risk job, knowing about a previous injury provides no information that will lead the worker to alter the assessed risk of injury. It is because high-risk jobs generate more injuries that injuries have informational content.

Although our general approach is based on a model of rational economic behavior, we try to explore the validity of that framework rather than to give results that are meaningful only if that model holds.

Figure 1.1 shows the components of the decisionmaking process related to the processing of information and subsequent behavior. The first set of concerns pertains to how individuals incorporate new information with their prior beliefs to form a revised assessment of the probability. In particular, how do individuals learn about risk, and can new information be introduced to influence this process?

For most decisions, prior beliefs concerning a risk may stem from a broad range of experience and related knowledge. For example, individuals' assessment of the likelihood of suffering a cut toe from a lawnmower will be based on their familiarity with the mechanical operation of the machine, their knowledge of the terrain to be mowed, their past experience in mowing the lawn (such as whether the mower tipped on hills), and their knowledge of lawnmower accidents to themselves or others. Adding a warning label to the machine may augment this knowledge, but it necessarily functions within the context of what individuals already know about lawnmowers. New information may alter but will not entirely correct excessively optimistic or pessimistic assessments.

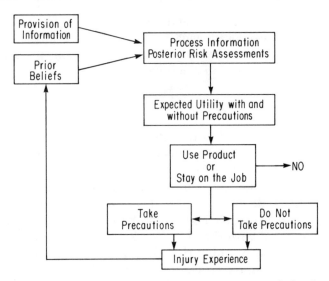

Figure 1.1 Information processing and economic behavior.

Evidence on individuals as probability assessors is at best mixed (for a review see Fischhoff and Beyth-Marom 1983; Tversky and Kahnemann 1974, 1985). At the most fundamental level, it is clear that individuals do not always behave as if they fully understand the laws of probability. After observing a long string of "tails" in the tossing of a fair coin, most people believe that a "heads" is due. At a more theoretical level, Arrow (1982) has speculated that risk perceptions in intertemporal markets may be particularly flawed.

In an influential study of biases in risk perception, Lichtenstein et al. (1978) addressed individual assessments of causes of death ranging from floods and botulism to heart disease. The overall pattern observed was that individuals overassessed the chances of low-probability events (such as tornadoes) and underestimated the chances of high-probability events (such as diabetes). Although these perceptions are clearly biased in a systematic manner, they do not necessarily reflect irrational behavior; they can be explained by the following adjustment process. Suppose individuals start by assessing the risk of all kinds of death as being identical. As they acquire information about each type of risk they will revise their beliefs toward the truth. If the information they have about the risks is not total, their revision toward the actual probabilities will be in the correct direction, but incomplete: low probabilities will be overassessed and high probabilities underassessed. A more appropriate test of rationality is whether the degree of learning in the correct direction is correlated with the level of the risk. Viscusi (1985) has shown that no such bias exists and that behavior is quite consistent with a Bayesian approach.

Much of the skepticism that has been expressed regarding individual rationality may stem from selecting the wrong basis for comparison. Some studies of individual rationality have used a model of perfect information as the reference point, rather than an optimal learning model in the presence of imperfect information. The latter approach is more relevant to the type of rationality to be expected for decisionmaking under uncertainty. Another limitation of previous research is its failure to provide any assurance that the experiments will produce behavior that will parallel the behavior observed in market contexts where individuals have a vested interest in the outcome and may learn through experience how best to utilize the information available to them. Although much of our research is also experimental, our studies utilized actual consumers and workers so as to recreate the types of contexts in which they make their decisions.

Until now, very little investigation of risk assessments has been directly related to economic behavior. Viscusi (1979) reported a strong correlation between the level of the industry risk and whether or not workers view their jobs as hazardous, but this evidence is only suggestive. To refine this test, in Chapter 6 we ascertain chemical workers' risk assessments using a continuous risk measure that can be compared with published risk statistics. Although the evidence for chemical workers provides particularly strong support for the accuracy of risk perceptions, these results in no way imply that all consumers and workers are fully cognizant of the risks they face.

In fact, our consumer study, which asked consumers about the relative riskiness of their households, provided far less reassuring results. Our approach there of posing the issue in relative risk terms is not unprecedented; the manner of framing risk issues has long been recognized as an important influence on survey responses. Informing drivers of the risk per trip of not wearing a seatbelt, for example, has a much smaller effect than if the risk is put on a lifetime basis (Slovic, Fischhoff, and Lichtenstein 1978).

Our emphasis in this book is not on individuals' prior risk beliefs per se, but rather on the role of these beliefs in conjunction with the addition of new information through labels. We investigate several aspects of the influence of information on behavior. One is whether or not risk information is provided. In the worker study (Chapter 6), we examined the key variables of interest, such as risk perceptions and quit intentions, both before and after the introduction of a chemical label. This longitudinal format greatly facilitates the analysis. Our approach in the consumer study was somewhat different. We divided the sample into groups, each of which was shown a different label. The differences among the groups resulting from the different labels formed the basis of the analysis. In both studies we will show how the presence of risk information alters behavior.

A closely related issue is the effect of the amount of risk information provided. In the consumer study, increases in the size of the warning label had the expected effect on inducing more precautionary behavior. In the worker study, it was possible to calculate the degree of information provided by the label relative to the amount of information the worker had initially, and this relative degree of information proved to be a pivotal determinant of a label's effectiveness.

This result highlights a major feature of both studies and a principle that is often ignored. The provision of new knowledge is more important to individual decisions than repetition of what is already known. An information program that provides no new information but simply exhorts individuals to alter their behavior is likely to be less effective than one that provides new knowledge in a convincing manner. Both surveys reported here focus on informational efforts rather than on policies of exhortation.

Although the amount of information provided by the label is of pivotal importance, how the information is provided greatly influences its ultimate effect on behavior. These concerns include the choice of color, of type size, of symbols, and organization. Chapter 6 explores the impact of different hazard warnings only through differences among labels for different chemicals. In the consumer study, format differences are a major part of the study design. Alternative labels for each particular product are included, and, most important, we explore whether reorganization of the information on risks and precautions improves the label's effectiveness.

Consideration of alternative label formats has a twofold purpose. First, from the standpoint of label design, it is desirable to know how labels can best be structured to be most effective. Improving the design of information transfer mechanisms depends on learning more about the relationship between a label's design and its effectiveness. Although the impact of label designs is influenced by cognitive processes, the task of designing an effective label is more of a technological question than a test of economic behavior. A second reason for exploring different label formats is to ensure that our results on how people process labeling information are not distorted by shortcomings in the label design.

After being given label information, individuals combine it with their prior beliefs to revise their assessment of the risk. (This stage is shown at the top of Figure 1.1.) If this learning process is fully rational, then it will satisfy Bayes' Theorem for altering probabilistic beliefs in a consistent manner. Evidence regarding whether or not individuals are Bayesians is somewhat mixed. Using cross-sectional data, Viscusi (1979) found that workers who have experienced a job injury, work in unpleasant conditions, or work in a hazardous industry are more likely to view their jobs as risky, as expected. Because the actual learning process was not observed, the evidence, though highly suggestive, does not necessarily imply that on-the-job learning takes place. The study of worker behavior in Chapter 6 investigates the learning process and the contribution of workers' prior beliefs and new information in influencing their risk assessment after seeing the warning label. Since the exact risks involved are not well known, the study tests the broad character of the learning process rather than whether there is exact conformance with Bayes' Theorem.

Psychological risk studies that have examined individual learning from information presented for hypothetical situations have found several systematic deviations from the behavior one would predict on the basis of Bayes' Theorem (for a review see Fischhoff and Beyth-Marom 1983; Tversky and Kahneman 1974, 1985). Two of these errors are that individuals give too much weight to the new information and exaggerate the completeness of their hypothesis sets. More generally, processing complex risk information may exceed individuals' cognitive abilities so that they rely upon heuristics to make these tasks more manageable.

The at least implicit recognition of these limitations by those engaged in providing risk information has greatly influenced the structure of actual information programs. The existence of limitations on human cognitive capabilities makes the format and wording of labels particularly important. In the absence of such limitations, it would be possible simply to give consumers detailed dose-response information about a risk and let them select the appropriate course of action. In contrast, the usual form of information transfer advises consumers of the kind of risks present (such as poisoning) and urges them in a directive manner to take a particular precaution.

Although there may not be exact conformance with a fully rational learning model, using a rational model for approaching economic behavior may generate good approximations of actual behavior. Moreover, it is not clear what weight should be placed on the existing evidence of irrationality. Consider the following oft-cited example from Tversky and Kahneman (1982, pp. 6–7): "Imagine an urn filled with balls, of which ⅔ are of one color and ⅓ of another. One individual has drawn 5 balls from an urn, and found that 4 were red and 1 was white. Another individual has drawn 20 balls and found that 12 were red and 8 were white. Which individual should feel more confident that the urn contains ⅔ red balls and ⅓ white balls, rather than the opposite?"

Most respondents in this situation incorrectly believe that the drawing of four red balls and one white ball provides stronger evidence that most of the balls are red. But how consequential are such examples of irrationality? Suppose the respondents had been given less information than in the example above and not been informed that the urn has a 50–50 chance of the described ⅔–⅓ mix of colored balls. Rather, they were told simply that the urn had some unspecified mixture of red and white balls. Then for many situations (depending on the respondent's prior beliefs) drawing four red balls and one white ball would have provided stronger evidence that most of the balls in the urn were red. Moreover, if the outcome of the second experiment is altered by a single ball—to eleven red balls and nine white balls—that draw would no longer be the stronger indicator of a predominance of red balls.

It is clearly unreasonable to assume that individuals can intuit all the laws of probability, which may involve considerable calculations even by probability experts. How much these shortcomings affect actual economic behavior under conditions of individual uncertainty is unclear. Our studies show that the general spirit of the learning process can be broadly characterized as Bayesian. The Bayesian expected utility framework is of value not only for its normative significance in suggesting how people should act, but also for its predictive power in many situations. Thus it is possible to view the Bayesian framework as providing a first approximation to actual behavior.

One cannot, however, dismiss all of the implications of studies contradicting the rational choice framework. The psychological risk literature indicates that there may be important and systematic errors in this learning process that may become particularly important when severe demands are placed on individuals' cognitive abilities. Thus, the departures from the model may be particularly great in certain experimental contexts. The extent of irrationality in observed market behavior has been less well documented. In this book we explore the limitations on individuals' rationality in the quasi-market contexts created in our two field studies.

Indeed, our results can be viewed as a partial response to Tversky and Kahneman's recent conclusion on the rationality of individual choices that "the claim that errors of judgment and choice will automatically be eliminated by learning . . . must be supported by evidence" (1985, p. 38). After reviewing the evidence on the systematic deviations of choice behavior from the assumption of rationality, they identify three possible defenses of the economic model of consumer choice, with learning behavior being one of them. Because in many choice situations effective learning is either slow or nonexistent, the relevant question for evaluating information programs such as hazard warnings is whether they induce behavioral changes in the desired direction and of sufficient magnitude to remedy the information problem. Ideally, hazard warning efforts will produce the same types of changes in risk perceptions that would occur through learning by experience, but at a more rapid rate and without the costs to the individual of accidents during the trial-and-error process.

Economic interest in individual learning is based on the impact of information processing on behavior. If no actions are altered, there is no economic value to the information. As Figure 1.1 illustrates, we assume that individuals first use the information to calculate the payoff, or expected utility associated with their actions. They then compare the expected utilities of alternative courses of action and pick the most highly valued alternative.

The principal feature of this approach is our assumption that individual choices are governed by the expected utility of alternative actions, that is, by the value attached to each outcome weighted by the assessed probability of its occurrence. Other approaches to decisions might involve minimization of risks or an attempt to avoid all substantial risks. As with the individual learning components, there is a substantial literature documenting deviations from the expected utility hypothesis in experimental situations.[1] A more limited body of literature, such as Kunreuther et al.'s (1978) analysis of flood insurance purchase decisions, questions the rationality assumption for actual economic behavior. Our focus is somewhat different: we use the expected utility model to estimate how much individuals value different health outcomes. In particular, what are their implied risk-dollar tradeoffs? We also address other components of the expected utility, such as the costs to consumers of taking precautions.

Whereas our results for workers' risk-dollar tradeoffs are of the same general magnitude as results in the literature, the consumer valuations of the low-probability events considered are implausibly large. This overreaction to low-probability events is a different type of irrationality from that encountered in the usual tests of the expected utility model, which typically focus on inconsistencies in attitudes across different sets of gambles.[2] The exaggerated response to low-probability events may explain why there is often a comparatively alarmist reaction to some very small health risks.

On the basis of the expected utility of alternative actions, individuals decide which activity they will undertake—whether or not to use the product, whether or not to remain on the job. We examine this fundamental choice in detail in Chapter 6, where we analyze how much workers will require to remain at their jobs, whether or not they intend to quit if their wages are not increased, and whether or not they would choose the same jobs if they had received the hazard information initially. The effect of both the level of the implied risk of the label and its informational content accords with an economic model of optimal experimentation.

The role of learning in risky situations has been developed in detail by Viscusi (1979), who includes a wide range of evidence indicating that workers in high-risk jobs are more likely to quit than those in low-risk jobs. Unlike the data from the worker survey in Chapter 6, the survey data used in Viscusi (1979) did not address changes in risk perceptions, so that the role of learning could be analyzed only through indirect tests. The results reported in Chapter 6 represent the first attempt to examine the effect on economic behavior of the actual learning process and the influence of the underlying determinants of risk perceptions, such as the informational content of the warning.

After making a product or job choice, individuals must select the degree of precaution that is desirable. Chapter 4 examines several consumer precautions, such as keeping products beyond the reach of children and wearing rubber gloves. In each case we find the expected effect of additional information on precautionary behavior. The results supporting the efficacy of information transfer efforts do not, however, imply full rationality. There is evidence of some imbalances in the precautions consumers select—overzealousness in some cases and comparative laxness in others.

The final link in the learning process at the bottom of Figure 1.1 is that injury experiences resulting from product use influence individuals' risk perceptions for subsequent behavior. Although the structure of both of our field studies does not permit examination of this form of learning, its importance has been well documented. As would be expected, on-the-job injuries are a major determinant of workers' risk perceptions (Viscusi 1979). A particularly intriguing study of learning by experience, involving dams, observed that the presence of a dam impedes individual protection against flood risks by preventing the small floods that otherwise would have provided valuable information about a future major flood (Burton, Yates, and White 1978).

Overall, a Bayesian decision model of economic behavior provides an instructive framework for structuring the process of acquiring information and using it to make decisions under uncertainty. From an economic standpoint, the Bayesian model rather than a model assuming full knowledge is the appropriate reference point for rational behavior in situations in which uncertainty is coupled with the opportunity for acquiring information. Although an optimal learning model has considerable appeal as a conceptual framework, if it is to be useful in predicting actual economic behavior, then the learning process should accord with its principal predictions. In the following chapters we explore the degree to which consumer and worker behavior is consistent with this framework.

The general economic issue that we address is how individuals respond to hazard information. Our first major result is that consumers and workers respond in the correct direction. The evidence in Chapter 4 indicates that consumer precautions increase with the amount of risk information, and the results in Chapter 6 indicate that chemical labels affect workers' risk perceptions, wage rates required to stay on the job, and quit intentions in the predicted manner. The risk-dollar tradeoffs underlying subsequent behavior are consistent with other evidence for job risks (Chapter 6), but they appear to be implausibly large for the very low-probability events faced by consumers (Chapter 5). Although the general direction of the economic response is rational, in some cases the extent of the response may reflect the limits on rationality.

The second major result of our studies is that it is possible to influence behavior through both the form and content of hazard information. In addition to providing principles for label design based on our knowledge of cognitive processes and consumer responses to advertising (Chapter 2), we show that the label format has an important effect on consumer behavior (Chapter 4) and that the degree of informational content of the warning is also instrumental (Chapters 4 and 6).

Some of these hypotheses are straightforward; others are not. One does not need a complex model to conclude that increased provision of risk information should increase individuals' incentives to exercise care. One hypothesis that is not readily apparent is the following. Using a model of optimal experimentation we can show that, for any given level of a risk, workers will demand more money to work on a hazardous job whose risks are well understood than they will to work on a similarly risky job whose risks are more speculative. The ability of rational models to predict such behavior contributes to our continued confidence in their usefulness in modeling economic behavior.

2 | Cognitive Considerations in Presenting Risk Information

JAMES R. BETTMAN,
JOHN W. PAYNE, and
RICHARD STAELIN

Designing an effective method for presenting information is an issue that has received considerable attention in the literature on the psychology of decisionmaking and marketing. In this chapter we review the findings in this literature, show how they relate to the provision of risk information, and then present guidelines for effective label design. These guidelines form the basis for the label design in the consumer study discussed in Chapters 3, 4, and 5.

Designing an effective information provision program is neither straightforward nor easy. The designer must be concerned not only with the content of the information but also with its format. There is extensive evidence from both basic and applied research that the same information presented in different formats can result in different decisions (for a review see Bettman 1979, Payne 1982). For example, in a study of consumer choices among supermarket products, Russo (1977) showed that very simple changes in the organization of unit-price information at the point of purchase resulted in shifts in purchasing patterns, affecting consumers' cost savings. Such findings make it essential that those responsible for the design of an information provision program thoroughly understand the cognitive processes of the target audience when that audience attends to and evaluates the information. In addition, by understanding how consumers (or workers) process information, designers can predict more accurately the effects of a particular format.

Figures 2.1 and 2.2 show two currently available bleach labels that use quite different formats to convey approximately the same safety and use information. Does one have a better format? Does it make any difference in behavior? If so, why is the format better? Are there any generalizable principles that would help designers decide if one of the two label formats should be adopted as a standard? Is there a still better format than these? To help answer these questions, we next review what is known about how people process information when thinking about risks.

2.1 Thinking about Risk: Human Processing Limitations

The bounded nature of rationality increases the need for careful design in programs to inform people about risks. Biases in risk perception can be traced to cognitive considerations, such as how easily a person can imagine or recall instances of a particular risk (Slovic, Fischhoff, and Lichtenstein 1982). Such a process for judging risks, often called availability, is valid in most instances, since more frequent events will generally be easier to recall, or their potential occurrence will be easier to imagine. However, ease of recall and imagination are related to other factors than just the statistical frequency with which events occur. For example, the same types of dramatic or sensational causes of death that people generally overestimate are those accidents that are heavily reported in the news media (Corules and Slovic 1979). Thus people are sometimes misled to perceive the wrong risks associated with an item.

There is also evidence that when individuals forecast certain natural hazards, such as floods, they are strongly conditioned by their immediate past (Kates 1982). That is, if an event has not happened recently, it is seen as not very likely to occur in the future. Reliance on personal experience may be particularly great for familiar hazards that are to some extent under the individual's control (Slovic, Fischhoff, and Lichtenstein 1980). Consider a household's use of toxic chemicals such as bleach. Unless the potential risks of such products are made vivid, people are likely to underestimate them, since the situations that are easiest to recall are those in which the product has caused *no* harm. This example is compatible

Figure 2.1 Clorox Bleach label.

Figure 2.2 Bright Bleach label.

with the finding that people tend to consider themselves relatively immune to common hazards (Svenson 1979; Rethans 1979; Svenson cited in Slovic, Fischhoff, and Lichtenstein 1980). Most people believe themselves to be better than average drivers, and the vast majority judge their ability to avoid accidents with common products to be average or above average.

Misperception of risk can have a significant impact on the success of any information provision program. Empirical evidence indicates that consumers will ignore information that they feel has little benefit. From an economic standpoint it is rational for individuals not to acquire information that has high costs relative to its benefits. Such behavior will influence the efficacy of information programs. If consumers perceive little risk (cost) associated with using a product, they are unlikely to seek out and process information about the potential risk in using the product. For example, Russo et al. (1986) have found that, unlike the provision of unit pricing information, the provision of nutritional information associated with proteins, minerals, and vitamins has little impact on consumer behavior. Although almost 50 percent of the consumers in their study became aware of the availability of this nutritional information and the information was presented in formats found in previous applications to facilitate use, only a small percentage found it desirable to take the time and effort to read through the available information. Even fewer seemed to use the information to alter their purchase behavior. On the other hand, when the nutritional information provided concerned negative attributes such as sodium, sugar, and calories, the information appeared to have significant impact (Muller 1982, Russo et al. 1986). The difference in the results of the two nutritional studies may stem at least in part from the fact that Muller's study provided information about nutrients, about which consumers perceive little risk (that is, from malnutrition), whereas Russo et al. provided information on attributes about which consumers perceive a much higher risk (such as salt and sugar content). This perception of risk, in turn, caused the consumers to seek out and use the newly provided information.

Risky choices pose particular difficulties for individual rationality. For example, choices posing both a gain and a loss raise different tradeoff problems (Payne 1985). Because of the difficulty of making decisions under risk, people often adopt heuristic techniques for trading off positively and negatively evaluated attributes. These simplifed choice rules normally do not use all the available information. The use of heuristics in making choices increases as the decisionmaking situation becomes more complex (Payne 1976).

A similar problem occurs when the amount of information provided is large. Such a situation is often called information overload. Although the empirical evidence is equivocal, there is some indication that making available large amounts of information can actually cause consumers to make poorer decisions (Malhotra 1982, Keller and Staelin 1985). For example, when comparing brands consumers might decide to focus solely on one item of information, such as price, if they perceive the number of brands or amount of information as excessive. Or they might examine one brand at a time to see if that brand has "acceptable" values on some subset of the brand's attributes. If it does, they may not make any further comparisons. Hence an important goal in the design of product labels is to present sufficient information for informed choices but not so much information that consumers will process it in suboptimal ways, which may lead to suboptimal choices.

An important principle governing the use of heuristics in complex decisionmaking situations is that individuals are adaptive. Many studies (for example, Payne 1982) have found that the strategies used are *contingent* upon the particular characteristics of the situation. Such effects are often called task effects, and properties of decisionmaking situations such as the number of alternatives, the format in which the information about alternatives is presented, and time pressure affect the heuristic techniques used by decisionmakers. This adaptivity to the situation is probably the most consistent finding in the research on decisionmaking.

The format in which information is presented is one task effect of particular interest. In Russo's (1977) study of unit pricing at supermarkets, mentioned earlier, the major comparison was not with no unit price information, but with the same information displayed differently on separate shelf tags. The improved format aided decisionmaking by *making the same information easier to process*. The study did not add either new alternatives or new information to the task environment of the supermarket shopper.

The importance of format is also borne out in a study of automobile safety by Slovic, Fischhoff, and Lichtenstein (1978). The basic hypothesis was that one reason people may not wear seat belts is the (correct) perception that the probability of a fatal accident on a single automobile trip is extremely small. A fatal accident occurs only about once every 3.5 million person trips, and a disabling injury only once every 100,000 person trips. Presenting risk information on a single-trip basis makes the reluctance to buckle up seem reasonable. The researchers reasoned that the probability of harm must be seen as exceeding some minimum value before people will respond to the risk. Consequently, they reformatted the information about automobile accident risks to a multiple-trip perspective. In a lifetime of driving, the probability of being killed is 0.01 and the probability of experiencing at least one disabling injury is 0.33. In an exploratory study, they found that expressed attitudes toward seatbelts (and airbags) were more favorable when the accident information was presented in the lifetime format rather than in the single-trip format.

2.2 Humans as Information Processors: The Structure of Memory

Thus far we have discussed a number of factors, such as perception of risk and format of information, that can alter a person's decision. We next show *why* these factors are influential, by describing some basic properties of the human mind as a processor of information. Over the past thirty years, psychologists have greatly expanded our knowledge of the human information processing system. What follows is a necessarily simplified overview of this system.

One of the most important theoretical postulates in current psychology is that people operate as information processing systems. Research has tried to describe behavior, such as a consumer's choice of a product, in terms of a small number of memories and processes (strategies) involving the acquisition, storage, retrieval, and utilization of information.[1]

The set of memories and processes that interact with the environment to produce behavior can be divided into three major subsystems: the perceptual system, the motor system, and the cognitive system (Card, Moran, and Newell 1983). The perceptual system consists of sensors (receptors) such as the eyes and ears and associated buffer memories. It translates sensations from the physical world (such as visual or aural input) into symbolic code that can be processed more fully by the cognitive system. The motor system, on the other hand, translates thought into action by activating patterns of voluntary muscles. Some of the research on the components of these two subsystems is relevant to the design of labels. For example, the amount of information a reader can take in with a single eye-fixation has been shown to be a joint function of the perceptual difficulty of the material (such as spacing of letters) and the skill of the reader. Furthermore, it has been estimated that the upper limit on the rate of reading, without ignoring some words in a text, is approximately six hundred words per minute. Such restrictions can become important in some instances (such as determining how much information to put in a television message). However, for the purposes of the design of effective labels, the most important subsystem to understand is the cognitive system.

In discussing the cognitive system, most researchers have found it useful to distinguish two types of memory: working memory and long-term memory (LTM). Working memory contains the information under current consideration. Long-term memory holds (stores) the individual's mass of available knowledge, including both facts and procedures. This distinction does not necessarily imply that there are two physically distinct memories. Working memory may simply be the currently activated portion of LTM. The crucial distinction is the difference in their functioning.

Working Memory

Working memory can combine both information from the environment, produced by the perceptual system, and information drawn (retrieved) from long-term memory. For example, in solving an arithmetic problem, a person uses both information (numbers) given in the environment and procedural information such as the rules of addition. Working memory also contains the intermediate products of thinking. Working memory is also often called short-term memory, reflecting the fact that items of information in working memory can be lost in twenty to thirty seconds if not actively rehearsed.

The most important fact about working memory is that it is limited in capacity; that is, only a few items of information can be considered at any one time. How few? The standard answer to this question is seven items, plus or minus two (Miller 1956), although more recently Simon (1974) has suggested that roughly four to five items is a more accurate estimate. This capacity limitation is easily demonstrated by a memory span task requiring recall of a sequence of items in correct order. For example, a person might be read the following twelve-letter sequence and then be asked to recall it in the correct order:

B-W-A-M-I-C-S-I-A-C-B-T

Most people would find this a very difficult task. Seven to nine letters is the limit for most of us. However, it is possible to increase the number of items of information recalled by recoding the information to form "chunks." A chunk might be best characterized as any piece of information that is represented as a single, meaningful item or that has some unitary representation in long-term memory. Consider a reordering of the twelve-letter sequence given above:

T-W-A-I-B-M-C-I-A-C-B-S

Most people in this culture can form the twelve letters into four chunks—TWA, IBM, CIA, CBS—that are easy to recall. This increase in recall through chunking can be dramatic. In one experiment a student was trained to recall eighty-one digits (Chase and Ericsson 1981). The student, an avid runner, was able to chunk the numbers into a much smaller set of items by relating the sequence of numbers to running times.

The capacity limitations of working memory have a number of important implications for decisionmaking behavior in general and for responses to informational labels in particular. For example, instances of information overload can be traced to limitations in working memory. There are limits to how much information it is reasonable to expect a consumer will be able to process from a label in any reasonable amount of time. Bettman and Kakkar (1977) have shown that people often do not transform the information, but instead process it in the form given. This is one of the reasons why the same information given in different formats (such as risk per trip or risk per lifetime) can have a different impact on a person's decision.

A second consequence of limited working memory is the use of heuristics to process information.[2] These procedures for systematically simplifying the search through the available information about a problem necessarily exclude some of the available information. Heuristics improve the chances of making a reasonably good decision given the limitations in processing capacity, while leaving some possibility of a "mistake." The use of heuristic strategies to solve problems and make decisions is one of the general principles of human information processing. Newell and Simon (1981) have argued that the use of heuristic search is at the heart of intelligence.

A major goal in the design of information systems is to take advantage of the power of heuristics while minimizing the potential for errors. Because most people will adopt simplifying strategies for processing the information on a label, effective labels are designed to encourage the use of such strategies. Viewing mental processing capacity as a scarce resource, designers of effective labels reduce the mental effort associated with the processing of information as much as possible so that people will tend to process more of the available information.

Long-Term Memory (LTM)

Unlike working memory, the capacity of LTM is generally thought of as infinite; for all practical purposes there are no limits to the amount of information that can be stored there. It has also been suggested that once information has been transferred from working memory into LTM it is never lost (Bettman 1979). Obviously, people do "forget" information; but it is suggested that forgetting really is just an inability to retrieve the information from LTM at a particular time. At other times, new retrieval cues or strategies may allow the person to remember the information.

Because of its capacity, LTM is sometimes viewed as external memory, just like a library, encyclopedia, or management information system (Simon 1981 elaborates on this view). Problem solving and decisionmaking would then involve a search for information in both the external perceptual environment and the memory environment, with information from one environment often guiding the search in the other.

Despite its unlimited capacity, not all information that is perceived (that is, placed in working memory) is transferred to or stored in LTM. One reason is the amount of time it takes to transfer an item of information to LTM. Writing to (storing an item of information in) LTM takes about seven seconds of processing effort. In contrast, Card, Moran, and Newell (1983) have estimated that retrieval (reading) from LTM is orders of magnitude faster than writing to it. According to these researchers, "this asymmetry puts great importance on the limited capacity of Working Memory, since it is not possible in tasks of short duration to transfer very much knowledge to Long-Term Memory as a working convenience" (1983, p. 41). This fact has implications for the sequence of operations that are likely to make up a decision strategy (Johnson and Payne 1985). It also helps explain why the organization of information for Russo's (1977) unit pricing study was so successful. It reduced

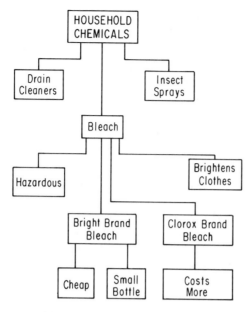

Figure 2.3 Possible encoding scheme for the concept of bleach.

the need to transfer information to LTM in making unit price comparisons among brands, because all the information was located externally on one list and the brands were listed in order of increasing unit price.

Capacity and read and write times to LTM are just some of the aspects of LTM that have been addressed in the literature. Two general issues of great importance to effective label design are how information is stored and how it is recalled.

The storage of information in LTM involves encoding operations. The most commonly discussed aspect of encoding is the representing of information in the form of semantic associations. That is, information is processed and encoded in the form of separate concepts and the associations among those concepts. Figure 2.3 provides an example of this form of representation for bleach with some encoded risk information. In this example the concept of risk is encoded with the generic concept of bleach instead of being associated with the two specific brands, Bright Bleach and Clorox. In this case then the consumer will not perceive different risk levels for the two brands. Consequently the riskiness of the product should not influence the consumer's decision in choosing between the two brands. If policymakers wanted consumers to choose among brands on the basis of risk, they would have to change the coding of the information so that the specific risk levels were associated with each brand.

How consumers encode information strongly influences how they ultimately use this stored information in making a decision. Several factors affect the encoding process. One important feature is that the acquisition of new knowledge (such as from a label) appears to be greatly facilitated by the existence of previously acquired relevant knowledge that can be used to form associations. This suggests, for example, that use of a common format and a common set of concepts in the labeling of hazardous chemicals would facilitate consumers' ability to encode hazard information about a new brand once the format had been learned through experience with other labels. In other words, the learned structure will enhance future encoding of new information that fits into the existing memory structure.

Knowledge must not only be encoded in LTM; because an individual acquires a vast amount of information, it must also be organized. Otherwise it would be impossible to retrieve a needed piece of information from LTM. Many psychologists think LTM is organized in hierarchical clusters of related knowledge. Figure 2.3 shows one such hierarchical form; Figure 2.4 portrays an expanded hierarchy for bleach, although it too is very simple. Furthermore, hierarchies are often embedded in other hierarchies. The bleach hierarchy, for example, could be part of a much larger hierarchy of cleaning agents.

The value of hierarchies in the recall of information is clear. Studies have shown that information learned in an organized, hierarchical fashion can be recalled much more effectively (for example, Reed 1982). Consider the memory span test discussed earlier. The ability to chunk letters into meaningful patterns, such as IBM or CBS, improved performance. However, trying to recall more than four or five such chunks is difficult for many people. Now consider the following sequence:

I-B-M-D-E-C-C-D-C-C-B-S-A-B-C-N-B-C

Possible chunks are IBM, DEC, CDC, CBS, ABC, NBC. The first three chunks represent computer companies and the second three chunks represent television networks. The overall hierarchy for both of these branches might be companies in the information/ communications business. Such hierarchies facilitate recall. One

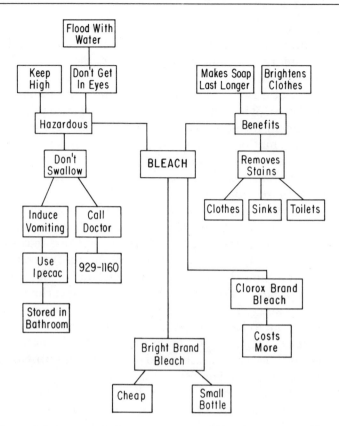

Figure 2.4 An extended semantic network for the concept of bleach.

of our later recommendations for effective design of labels builds on the implications of hierarchical structure research.

2.3 Policy Implications of Consumer Processing Limitations

The summary above has indicated that consumers have limited capacity to process information and that coping with information about risks is a complex and difficult task. These conclusions are contrary to the typical policy assumptions that consumers are extensive information processors and that providing more information is always helpful. In contrast, consumers may in fact use simplifying heuristics to limit processing. Thus merely making information available may not be sufficient. Instead designers must distinguish between the *availability* and *processability* of information (Russo, Krieser, and Miyashita 1975). Processability refers to the ease with which information can be comprehended and used. In general, information must be both available and easily processable to be utilized.

The processability of information is a function of how the information is presented. That is, presenting information that is well organized and in formats that facilitate processing can increase use of that information. For example, providing information about potential hazards from poisoning on a common rating scale for products containing toxic chemicals would facilitate risk comparisons among those products.

Processability is also a function of the kind of processing encouraged or required. Consumers may use different processing strategies depending upon the task; but policymakers may wish to encourage certain types of processing. For example, when the consumer is purchasing the product, the policymaker may wish to encourage comparisons of potential hazard among various brands. On the other hand, when attempting to use the product, the consumer would like to have simple instructions so as to avoid risks. Different types of formats and organizations of information facilitate different types of processing. No one format is optimal for all types of information and/or situations. Rather, processability depends upon the *congruence* between the format and organization of the information and the type of processing to be done (Bettman 1979).

Effects of Format and Organization

There are two basic approaches to congruence. The first is reactive. That is, label designers can attempt to determine how consumers are currently processing information and utilize formats that will make that type of processing easier. A second approach, particularly relevant for policy, is more proactive. Policymakers can determine types of processing they *wish to encourage* (such as making more comparisons among brands) and design formats that facilitate such processing. Because some evidence indicates that consumers tend to process information in the format in which it is provided rather than transforming it, policymakers may be able to facilitate certain types of processing through judicious design of information provision.

Three major considerations govern general principles of format and organization as they relate to processing tasks:

1. Reducing the cognitive effort and/or time needed to locate the externally available information, retrieve any previously stored information, and encode the newly provided information

2. Reducing the cognitive effort and/or time needed to make risk-benefit tradeoffs within a particular brand or alternative being considered

3. Reducing the cognitive effort and/or time needed to make comparisons among different brands or alternatives

One important goal in designing labels is to make it easy for the consumer to locate particular pieces of information when needed and to encode their meaning once located. For example, information on how to avoid certain risks and on appropriate antidotes should be easy to find; the consumer should not have to search through fine print or hunt all over the label. When found, the information should be easily understandable.

Several design principles can be used to facilitate ease of location and encoding. One is to make the information salient by using different colors or sizes of type (such as large letters in a color that contrasts with the other printing on the label). Even more crucial is consistent organization. If information on various factors was put in the same relative position on all labels—that is, if the consumer knew that antidote information was always at the bottom of the label or that instructions for avoiding risk during use were on the middle-right of the label—locating the desired information would be much easier. In addition, as noted in the earlier discussion of memory, processing will be facilitated if such a common format is arranged in a way that is consistent with how people like to process and encode the information. Since hierarchical organizations seem very effective, information may be more readily processed if it is presented hierarchically in the order in which consumers are likely to use it. Kanouse and Hayes-Roth (1980) developed the following hierarchy for prescription drugs: what the product is, what its benefits and risks are, how it should be used, what risks in use there are and how these risks can be avoided, and what should be done if the product is not used properly.

To facilitate the encoding of information once it is located, the information should be simple and easily understood. One device that might be very useful is the use of symbols wherever possible. For example, symbols could be used to represent the degree of various potential hazards. Figure 2.5 shows hazard symbols used in Canada. By using symbols that immediately connote specific types of hazards, this system depicts both the type and degree of hazard in a readily understandable fashion.

Thus there are four general ways to facilitate ease of location and encoding information:

    1. Make important information more salient by means of color and/or type size.

    2. Use a common organization for information on all labels.

    3. Design this common organization hierarchically and in a manner compatible with the scheme used by most consumers to store information about the product.

    4. When possible, use symbols which quickly convey the concept.

Policymakers also may want to encourage consumers to make risk-benefit tradeoffs within a given alternative by getting them to

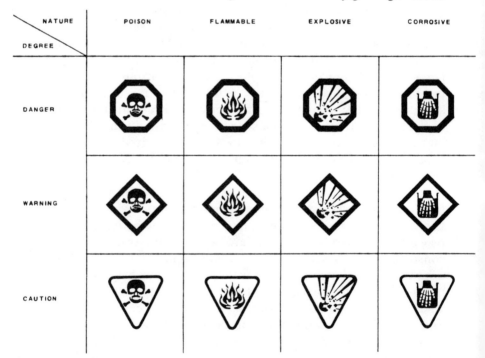

Figure 2.5 Canadian hazard symbols. (Source: Canadian Hazardous Products Act of 1970.)

Figure 2.6 Sweden's product efficacy label. Arrow shows average performance level of this brand/product; heavy bar shows performance range of all other brands/models tested. (Source: Hans Thorelli and Sarah Thorelli, *Consumer Information Handbook: Europe and North America* [New York: Praeger, 1974], p. 48.)

compare the risks and benefits for a specific product before deciding whether or not to purchase it. Presenting information on the benefits of the product in one place and information on its risks in another place, preferably close together, will facilitate this process. The following three steps are involved:

    5. Collect information on benefits in one place on the label.
    6. Collect information on risks in one place on the label.
    7. Organize the label so that the information on benefits and risks is in close proximity.

Finally, a label should facilitate comparisons among several alternatives. For example, the consumer may wish to find the product for a particular use with the least hazard from poisoning. Providing information on a common hazard scale and with the same overall organization on each label will help. However, this format still requires that the consumer examine several different packages. One way to communicate comparative information would be to provide on each product's label the range of potential hazards over all products in the category. Figure 2.6 shows an example. A similar approach is already used on many household appliances and automobiles: the minimum and maximum energy consumption of any brand within the product class is given as well as the particular brand's energy consumption.

Unfortunately, even such comparative scales do not always make comparisons easy enough. Knowing that there are products with lower hazard levels does not help the consumer find them. For this reason product labels alone may not provide enough information for comparisons among products. As in the unit pricing experiments, a list of products ranked by potential hazards and affixed to the shelf at the point of purchase could facilitate consumers' task. To facilitate comparisons among brands, we recommend two activities:

8. Provide information in a relative or comparative format.

9. Consider using in-store comparative lists in addition to labels.

Task and Information Presentation Modes

Before applying these broad guidelines to the design of labels for products containing chemicals, we briefly consider how different processing tasks and different modes of presenting information interact with these principles.

The two general types of situations faced by consumers regarding products containing hazardous chemicals are purchase and use. In the purchase situation, facilitating risk-benefit tradeoffs for a specific brand and comparisons among brands is particularly important. The consumer may make a buy/not buy decision for a whole category of products based upon tradeoffs of benefits and risks. If a decision is made to purchase, comparisons among brands may be crucial. For the usage situation, on the other hand, ease in finding information on proper usage and on antidotes in case of an accident is more relevant. Clearly, different types of information assume different importance in different situations, and policymakers should take this fact into account when deciding how information is to be provided.

Different modes of providing information also have different properties. For example, television or radio messages allow limited time for processing. Hence, because of the limitations on working memory discussed earlier, complex information probably should not be presented in such messages. Broad notions of benefits and the potential for hazard are probably all that can be readily communicated. Print advertisements can provide more information, since the reader can control the speed at which the information is received. However, print ads, like TV and radio messages, are generally not available at the point of purchase. If the information is to have an impact there, consumers must recall the potentially complex information from long-term memory, a task that often is difficult.

Labels, however, are available at the point of purchase, and judicious design can make risk-benefit tradeoffs easier and provide clear use instructions. An in-store list can facilitate comparisons among brands. Package inserts can provide even more detailed information. However, reliance on inserts cannot be primary, because they are usually not easily accessible at the time of purchase and because they may be discarded and thus be unavailable during use, particularly if the product is used on multiple occasions.

Because no one method is effective for all types of processing, an *information provision system* is necessary, using several methods that complement one another.[3] The following section describes such a system for products containing hazardous chemicals. Although the label plays a central role, it is not the only component.

2.4 A Labeling System for Products Containing Toxic Chemicals

A labeling system provides information to consumers in many ways (such as ads, in-store displays, package labels, package inserts, and general consumer education programs). Each of these ways is best suited for providing certain types of information but may be ill suited for other types. For example, in-store displays may be very effective in helping consumers compare different products that could accomplish the same purpose. However, such displays would not be as effective for communicating safe usage instructions at the time of use. Hence a labeling system has a number of advantages. First, it can provide information in several different formats, each one tailored to the particular information processing situation. Second, since not all the information has to be in one place, the information provided in one part of the system can complement that provided in other parts of the system.

Our labeling system has four major components: advertisements, point-of-purchase (POP) displays, labels, and package inserts. Although we concentrate on the labeling portion, all parts of the system play important roles in providing information to consumers. We did not include general consumer education programs, because the usual objective of such programs is to provide general knowledge about a product class or product classes rather than specific knowledge about one or more brands within a product class. Moreover, there is little evidence that these programs are cost effective in terms of modifying consumer behavior (see especially Staelin 1978 and Adler and Pittle 1984). Consequently, although education programs that provided consumers with generalized knowledge or with heuristic strategies for better evaluating the risk aspects of a product class or the risk inherent in using that product class would be helpful, our proposed labeling system need not rely on the existence of such programs.

Our proposed labeling system has five goals: (1) *motivating* the consumer to examine information about the toxic chemicals, both at time of purchase and at time of product use; (2) making it easy for the consumer to understand the *level of hazard* potentially posed by a particular product, both at time of purchase and at time of use; (3) making it easy for the consumer to *compare at the point of purchase the levels of hazard* posed by several products that could be used for the same application; (4) making it easy for the consumer to see how to *use the product safely;* and (5) enabling the user to take the *appropriate action if the product is misused.* Thus, we want to increase the likelihood that consumers will have available and will use the appropriate information both for deciding whether or not even to purchase the product (level of hazard and comparison of such levels) and for learning how to use the product safely once it has been purchased. The components of the labeling system vary in their effectiveness in achieving each of these goals.

Advertisements

Ads convey information to consumers at a different time from either purchase or use. To use the information at these other times the consumer must encode the message and then retrieve it later. Ads, then, are not in general as effective in providing information as POP displays, labels, or package inserts, which are available at purchase or use. Moreover, because TV ads are short (normally thirty seconds or less), they are unlikely to convey detailed safety information effectively.

This does not mean that ads cannot be used to enhance a labeling system. Ads can effectively convey general moods and settings. Also, some research indicates potential benefits from ads that show consumers looking at and using labeling information (Wright 1979). Thus, instead of requiring the manufacturer to provide specific label information in an ad, as the Federal Trade Commission does, it might be better for the Environmental Protection Agency (EPA) to require manufacturers to show consumers reading labeling information, with a brief verbal statement that such products can be hazardous and that one should always consult the label.

The purpose of providing the correct "model" behavior and a brief warning in the ad is to *motivate* consumers to do the same in the store. Similarly, ads that model proper use of the product, such as showing people wearing rubber gloves when using the product,

RISKS

BRAND	Swallowing	Contact	Breathing	Flammability
Drāno	XX	X	XX	——
Liquid–Plumr	XXXX	XX	——	——
Brand C	XX	XX	——	——
Brand D	XX	XX	XX	——
etc.				

Figure 2.7 Rating the risks: A simple scheme for drain openers. The more symbols, the greater the risks.

may also be of value. In other words, the main goal of ads with respect to hazard information should be to motivate the consumer to consult other components of the labeling system rather than to deliver detailed hazard information. In this sense ads could be thought of as an educational program, in that they provide general strategies and knowledge about a product or product class rather than detailed product information.

Point-of-Purchase Displays

Point-of-purchase (POP) displays have been shown to have a strong influence on purchase behavior in situations in which comparative product information has value (Russo, Krieser, and Miyashita 1975; Russo 1977; Wright 1979). Thus they are appropriate when it is desirable to shift usage patterns so that the least hazardous product types are used for any given application. Determining these least hazardous product types requires comparisons *among* products. Such comparisons may not be easy to make, even if labels are well designed. Moreover, if comparisons are difficult, it is much less likely that they will be made.

Perhaps the best examples of the impact of comparative product information are Russo's (1977) studies on the use of unit price and nutritional information. An analogous POP display would be a list of products ranked by degree of potential hazard. Such lists would be most valuable for products with varying degrees of hazard that can be used for the same application.

To implement the POP component, retailers would post lists of hazard levels for specific hazards for all brands that claim they can be used for a specific application. Figure 2.7 shows a list of products designed to unclog drains. The list might show dry chemicals (such as Drāno) and liquid chemicals (such as Liquid-plumr). Since these two types of products pose different types and levels of risk, the comparative information would make it easier for consumers to trade off risks and benefits. Such a display would not only increase consumers' ability to compare several types of products, but would also make consumers aware of the total set of alternatives for a particular application.

Three classes of practical tasks are involved in implementing this component of the system. The first, specifying all the products that can be used for a specific purpose, might be difficult. The second, collecting the information on risk levels for each product on all hazard dimensions, would not be too onerous if our proposed labeling system were implemented, for the quantification of risk levels would appear on each product's package label. Third, preparing and maintaining the lists would involve difficulties for either the retailer or the government. Requiring retailers, particularly small ones, to provide a list of products available in that outlet would impose a high cost burden. Such costs would presumably then be passed on, at least in part, to consumers. On the other hand, lists prepared by the government would need to include all products, whether stocked in that store or not, and such preparation might be difficult and costly. Although these cost issues are significant, empircal evidence indicates that the potential benefits also can be substantial in situations in which the information presented is valued by the consumer (Russo 1977). Hence this cost-benefit tradeoff should be seriously considered.

Labeling

Because the label provides information both at time of purchase and at time of use, it has the potential to meet two of our goals: communicating level of hazard and telling the consumer how to use the product safely. In fulfilling these goals it can affect whether or not a product is even purchased and how it is used. Therefore, we have tried to develop a format for the label that maximizes its usefulness in both situations. In addition, labels can be used to cue the use of other sources of information. Ley (1980) has argued that warning labels are unlikely ever to be adequate as the sole source of information on health risks. He suggests that people want more detailed information on why a product is harmful and on what to do than can be provided by a label. Consequently, he proposes that the warning label act as a cue for recall of more detailed information provided by, say, package inserts. This idea of referencing other sources fits into our concept of a total information system.

Our goal was to develop a single label design that would provide useful information to both experienced and inexperienced consumers. We did this by applying the general principles outlined earlier in this chapter. Figures 2.8 and 2.9 show two labels designed in accordance with these principles.

All labels should have seven components: (1) a statement of what the product is (name and ingredients), (2) a statement of the product's benefits (uses), (3) a symbolic visual display of risk levels for specified types of hazards, (4) a statement of the types of bad outcomes (dangers) that can occur, (5) a description of the actions that should be taken to avoid such dangers, (6) a statement of how the product should be used to derive the stated benefits, and (7) a statement of the antidotes (actions) that should be taken if a bad outcome occurs. These components allow the consumer to trade off risks and benefits and provide information relevant to safe use.

We specified a standard format for delivering this information because we believe this will facilitate consumers' abilities to use and encode it. Thus if a particular type of information is in the same place on all labels, consumers will quickly learn how to locate that information rapidly (see also Hadden 1985).

In addition to having a common format, the label is organized so as to facilitate processing. Thus our label is designed hierarchically, with the information listed in the order in which consumers might use it. The general order and organization are as follows:

Name of the product

Benefits   Potential hazards

Dangers (bad outcomes)   How to avoid dangers

How to use to derive benefits

Antidotes

The two labels in Figures 2.8 and 2.9 follow this organization. Ingredients might be listed on the "front" part of the label. The information above largely concerns the "back" part of the label.

In addition to this hierarchical ordering, the organization of the label attempts to facilitate processing in other ways. Benefits and risk information are each collected in one place, and the two types of information are put next to each other so that consumers can more easily make the tradeoff between risks and benefits at the

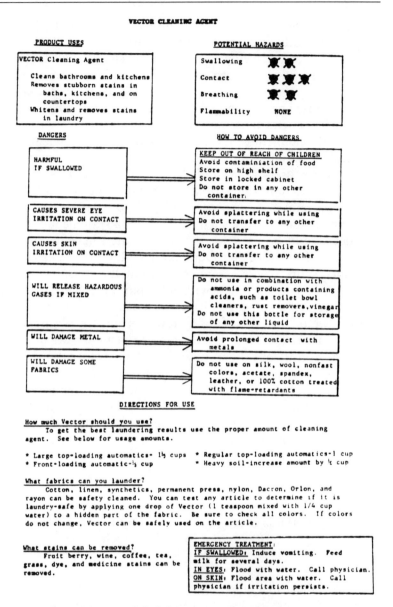

Figure 2.8 Test label for Vector Cleaning Agent.

point of purchase. Making such tradeoffs easier could affect the purchase rates of generally hazardous products.

We have also graphically linked the possible dangers in use with the recommended actions to avoid these dangers, using boxes and

UNSTOP LIQUID DRAIN OPENER
READ ENTIRE LABEL BEFORE OPENING OR USE

PRODUCT USES   POTENTIAL HAZARDS

UNSTOP Liquid Drain Opener

Opens clogged drains
Keeps drains open
Will not harm pipes
  or septic tanks

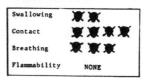

Swallowing

Contact

Breathing

Flammability   NONE

DANGERS   HOW TO AVOID DANGERS

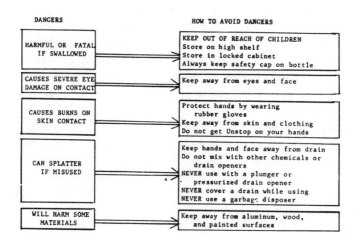

HARMFUL OR FATAL IF SWALLOWED	KEEP OUT OF REACH OF CHILDREN Store on high shelf Store in locked cabinet Always keep safety cap on bottle
CAUSES SEVERE EYE DAMAGE ON CONTACT	Keep away from eyes and face
CAUSES BURNS ON SKIN CONTACT	Protect hands by wearing rubber gloves Keep away from skin and clothing Do not get Unstop on your hands

CAN SPLATTER IF MISUSED	Keep hands and face away from drain
	Do not mix with other chemicals or
	  drain openers
	NEVER use with a plunger or
	  pressurized drain opener
	NEVER cover a drain while using
	NEVER use a garbage disposer

WILL HARM SOME MATERIALS	Keep away from aluminum, wood, and painted surfaces

DIRECTIONS FOR USE

TO OPEN: Place bottle on flat, steady surface. Press down on cap while turning in a counter-clockwise direction. Do not squeeze bottle.
TO LOCK: Turn cap onto threads in a clockwise direction until it no longer turns. Then press down cap and turn to seal.

KEEP DRAINS OPEN: Every week remove drain sieve. Keep bottle away from face at all times. Use 1/4 bottle. Let stand 10 minutes. Flush with hot water from faucet.

OPENING CLOGGED DRAINS: Remove sieve and any standing water from sink, wash-bowl, or tub. Keep bottle away from face at all times. Use 1/2 bottle. Allow to work 30 minutes. Flush with hot water when drain clears. Repeat if necessary. For tough jobs, let stand overnight before flushing with hot water. Keep face away from drain.

44600 00102

EMERGENCY TREATMENT:
IF SWALLOWED: Do NOT induce vomiting. Give large quantities of water or milk. Seek medical care immediately.
IN EYES: Immediately flush with water for at least 15 minutes. Seek medical care immediately.
ON SKIN: Flood area with water for at least 15 minutes. If irritation persists, seek medical care.

Figure 2.9 Test label for Unstop Drain Opener.

arrows. For example, the label in Figure 2.8 both tells consumers that the product could release hazardous gases if mixed and gives specific instructions on how to avoid such a hazard. This specific, concrete information is formatted to maximize its ease of application at the time of product use. We feel that it is crucial to have very clear usage guidelines on the label itself, which almost always is available when the consumer tries to use the product. Finally, we have placed the antidotes at the bottom of the label because we believe (less strongly) that this is where most consumers first look for such information. Even if this is not currently the place consulted first, if our labeling standard is adopted consumers will soon learn where the information is located. The abundance of information on the label reflects our belief that some redundancy is beneficial.

In addition to graphic devices (arrows and boxes) to organize information and to promote direct processing, we have used symbols to communicate risk hazard levels. Although it would have been possible to use more extensive symbolic systems such as those shown in Figure 2.6, we felt that use of the familiar skull and crossbones sign for danger represented a reasonable initial test system. More details could always be added if necessary.

These design considerations implement the general principles 1–7, discussed earlier. However, principle 8, providing information in a relative or comparative format, has not been addressed. Because we feel that it is highly desirable to show the range of hazard for other products suitable for the same application, we would like to communicate something like: "This product has level 4 of risk on contact. Other available products that accomplish the same application range from 2 to 4 on this hazard." Figure 2.6 provides one example of a scale designed to do this; similar systems are used for automobile mileage ratings and energy usage. We would like to provide such comparative ratings symbolically for risks, but we are unsure whether it is feasible (because a single product often has multiple applications) or how best to do it. Such comparative information is better provided in a point-of-purchase display. However, if such displays are not used, further consideration should be directed to how best to provide comparative information on the label.

Implicit in the labels shown in Figures 2.8 and 2.9 are a list of hazards and a rating scheme for scaling the hazards. In these examples we have shown four hazards—from swallowing (ingestion), contact, breathing (inhalation), and flammability—and have used a scale with five degrees of hazard (zero to four skull and crossbones symbols, with more symbols implying greater hazard). However, determinations of the hazards to include and the scale values are not within our range of expertise. Technical authorities would need to be consulted regarding measurements and other facets involved in the design of a rating system.

Even with these caveats about the symbolic hazard section of the label, we feel strongly that it is important to present hazard levels symbolically on specific dimensions. In developing this scheme for presenting hazard (risk) information, we decided to concentrate on the severity of consequences if the product is misused (for example, what happens if it is ingested). Thus we do not consider remaining risks if the product is used safely or the probabilities that the various hazards will occur. We had two reasons for this approach. First, we felt that the best way to protect consumers from harm was to notify them of *potential* hazards. Knowing that a product can be dangerous if used improperly should motivate more careful use. Our label does not communicate risks that remain even with proper use, since government agencies usually adopt bans when these risks are substantial. Second, we focused on severity and provided no frequency information because humans have difficulty processing probabilistic information and individuals tend to focus on severity as long as probabilities are above some minimum cutoff level. Only hazards whose probabilities are above that minimum should be included on the label. Perhaps package inserts could provide probability levels.

Finally, we have handled effects on special populations implicitly rather than explicitly. If the product is hazardous to a special population group that is significantly large, the hazard rating on the label should reflect the level of hazard *to that population* rather than to the average user. This approach is consistent with our emphasis on communicating the most serious potential hazard. Hence we would present the hazard level (and possibly mention particular populations on the label) only if the subpopulation were reasonably large (for example, children) or if the dangers were especially severe for this subpopulation (for example, a potentially fatal allergic reaction).

In dealing with long-term effects of chemicals, such as carcinogenicity, we were less certain how to proceed. We decided not to include long-term effects in our list of hazards because severe measurement problems are associated with such a dimension. However, if the regulatory agencies can measure long-term effects and are willing to rate each chemical on this dimension, then they should be included in our symbolic hazard section.

A second alternative for handling long-term risks is to include a brief verbal statement such as the one included on cigarette packages. Such a statement would probably be less effective than our symbolic rating scheme because of the difficulty of rating the long-term effects of a product on a person's health. If the regulatory agency believes verbal statements are the only feasible way politically and/or technically of quantifying long-term risks, we would recommend that such information be included in the hazard section.

A third option, and the one we implicitly assumed in our examples, is that any such long-term effects are incorporated in the basic rating scheme for each hazard shown on the label.

In summary, we feel that long-term hazards are really a measurement issue that requires technical expertise and are not a design issue. Our label can display the long-term hazard information if it can be quantified.

Finally, our label requires each producer to allocate a substantial amount of label space for hazard information. Because label space is often scarce, our standard format might exclude other information, such as product benefits or potential uses. We have analyzed the space allocation of our labels for a liquid drain cleaner and a bleach. The total amount of label space (total area of label) for our bleach label and for other actual labels (such as those shown in Figures 2.1 and 2.2) was approximately the same. However, we devote a greater share of that space to risk information—69 percent, versus 41 percent on the Bright Bleach label and 31 percent on the Clorox Bleach label. On the other hand, our label for drain openers uses 63 percent of the available space for risk information versus 78 percent on an actual label. Thus our label will not always require more space allocation for risk information.

Package Inserts
Package inserts can be used to provide detailed information at the time of use. Because the insert is not as constrained by space limitations as the package label, it can give more detailed information. However, because inserts can be lost, particularly in the case of products that are used multiple times, the label must be the major component for communicating essential usage instructions. The package insert could provide details on how best to gain the *benefits* of the product, with the label providing information on how to

avoid the *risks*. The insert could also provide more details on the risks and on usage instructions. However, everything essential about risks and proper use should be communicated on the label.

2.5 Conclusion

The labeling system described above is an example of a system for providing information based upon an understanding of human information processing limitations and characteristics. Such considerations are exceptionally important in the design of labels and other components of the system. Information must be made not only available but also processable. If processability is ignored, information about hazards may not be utilized, with substantial negative consequences. Making risk information more easily processable will result in societal benefits that can outweigh the costs of providing such information.

3 | The Design of the Consumer Information Study

WESLEY A. MAGAT,
W. KIP VISCUSI, and
JOEL HUBER

3.1 The Consumer Decision Problem

To analyze how consumers process risk information, we designed
a survey that would address the effects of the labeling of potentially
hazardous products. Labels on products such as pesticides and
household cleaners provide not only instructions and a description
of product uses, but also information about the hazards associated
with using each product and the precautions needed to avoid those
hazards. On the basis of this information and any prior knowledge
about the efficacy and risks associated with a product (and other
substitutes), the consumer must first decide whether to purchase
the product, then how often and in what ways to use it, and finally
what precautions to take before, during, and after use.[1] To restrict
the complexity of the problem, we assumed that the product had
been purchased and the decision had been made to use the product,
leaving the consumer with the choice of which precautions to take.

From an economic standpoint, consumers' precaution-taking decisions will be governed by both the costs and the benefits of each
possible precaution, such as wearing gloves, storing the product

in a childproof location, or avoiding breathing fumes. Consumers take precautions whose benefits exceed their costs, and they forgo the other precautions. The costs consist of the disutility, time, and mental effort required to take the precaution. Although these costs may be small for each application of a product, such as a household cleaner, over an extended period, such as a year, the costs can be significant for an often-used product. The benefits of precaution-taking consist of the reductions in risk associated with the product, such as hand burns, child poisoning from ingestion, and injured lungs. To estimate the benefits of precautions the consumer must assess both the effect of taking the precaution on the probability of each possible injury that it protects against and the value of the resulting reduction in the risk of injury.

The consumer survey addresses several aspects of the link between labeling information and precautionary behavior. The first concern is the nature of the information provided, both the amount of risk information and the manner in which it is conveyed (that is, the label format). We address the effect of each set of information on precautionary behavior. The main determinants of precautionary actions are consumers' risk perceptions, the disutility of precautions, and valuations of the injury risk reductions. The findings regarding consumers' risk-dollar tradeoffs reveal major inadequacies in consumers' responses to low-probability events. Although we ascertain these preferences by assessing consumers' tradeoffs between product price and product risk, our focus on the market implications of risk information is less extensive than in the worker survey results in Chapter 6.

3.2 Product Selection

To study the consumer decisionmaking problem posed by product labeling, we chose two products and monitored consumers' responses to the information on their labels in a laboratory setting. We placed several different labels on each product, randomly assigning consumers to labels and presenting each consumer with only one product and label. The labels were differentiated by the amount of information about hazards and precautions and by the format in which that information was presented.

The questionnaire was designed to elicit several different types of information. First, it measured consumer intentions to take several different precautions associated with the use of the product.[2] Second, it obtained consumer valuations of the disutility of taking those precautions. Third, it sought their valuations of avoiding the various possible injuries associated with using the product. Finally, the questionnaire collected pertinent demographic information about the subjects.

There are three broad classes of hazardous chemical products that we could have used in the study: household, industrial, and agricultural. Constraints on the number of subjects available and on ease of access to those subjects forced us to focus on a single class of products. We chose the household chemical class and selected two representative products: a liquid cleaning agent and a liquid drain opener. Because responses to the labels on products in the three different classes might well differ, we decided to explore the consistency of responses to two products within the same class, rather than drawing conclusions based on only one product from a class.

Since our methodology for assessing the value of information (described below) is more appropriate for identifying short-term changes rather than long-term changes in behavior, we selected chemical products for which label changes would cause short-term changes in precautionary behavior. This focus on short-term responses required the use of new and unfamiliar products to encourage consumers to read the labels and learn about the necessary precautions from those labels rather than relying upon experience with similar products.

For this reason we called our first product a cleaning agent, even though its use and hazards were equivalent to those of common household bleach. Similarly, our liquid drain opener product was actually a hybrid between a conventional liquid drain opener and a granular drain opener containing 100 percent lye, which has potentially more serious health effects than the liquid formulation. Changing the label on a well-known product, such as bleach, would produce little short-term change in precautionary behavior because most consumers already know the hazards of the product and what precautions are necessary to use it safely. This conjecture was strongly supported by evidence from some of our pretesting, in which we called the product bleach. In contrast, the labels on new products have much more influence on the extent of precaution-taking by product users, especially in the short run.

The products were also chosen to be representative of those that are or could be regulated by EPA's labeling rules. Labels for bleach products that make biocidal claims are already regulated by EPA's

Office of Pesticide Programs, and drain opener is similar to other household chemicals that in the future may be regulated by the Office of Toxic Substances under the Toxic Substances Control Act. We also chose drain opener and bleach because they are widely used products that cause significant numbers of injuries every year in the United States. Thus, despite the low accident rates per bottle used, the risks examined would be of sufficient magnitude to generate precautionary behavior. We also sought products that create short-term hazards rather than long-term ones, such as cancer or birth defects, because short-term risks are easier to communicate accurately. Finally, for both products there are easily identifiable precautions to be taken when using them. A label that directs a consumer to wear rubber gloves has a more readily testable effect than, say, a label that urges a consumer to exercise care when using a product.

Table 3.1 shows two of the primary hazards associated with each of the products studied—chloramine gassing and child poisonings for cleaning agent, and hand burns and child poisonings for drain opener. Although several other hazards are identified with the unsafe use of both products, the Consumer Product Safety Commission data on product-related injuries reported by hospital emergency units and data from the National Clearinghouse for Poison Control Centers indicate that the two hazards associated with each product cause a significant portion of the injuries associated with their use.

Because most readers may not be familiar with the four injuries and how they arise, we briefly describe them here. Mixing bleach with ammonia and ammonia-based products, such as toilet bowl cleaner, causes chloramine gas to form. Inhalation of this gas causes headaches and burning lungs, eyes, and nose and may require the victim to be hospitalized for several days. Excluding drug overdoses and other poisonings of a "voluntary" nature, chloramine gas poisonings are the leading cause of poisonings among adults.

Table 3.1 Hazards and precautions associated with study products

Product	Hazard	Precaution
Cleaning agent	chloramine gas	do not mix
	child poisoning	store in safe place
Drain opener	hand burns	wear gloves
	child poisoning	store in safe place

The child poisoning risk is more direct. If young children drink bleach accidentally, they may have difficulty breathing, may vomit, and may complain of stomachaches.

The first hazard listed for drain opener is hand burns. Splashing liquid drain opener on the hands can cause painful burns or red, swollen blisters. Complete healing usually occurs within a week. A much more serious outcome occurs when children drink liquid drain opener. The resulting severe and painful burns to the mouth and throat may require an operation to replace parts of the throat and hospital treatment for up to three weeks. In extreme cases, the child may permanently lose the use of the esophagus.

Table 3.1 also lists the precautions associated with each of the four hazards. Consumers can avoid chloramine gas poisoning by refraining from mixing bleach with other ammonia-based products. They can prevent hand burns from liquid drain opener by wearing gloves. Finally, both types of child poisoning can be avoided if the product is stored in a safe place, such as on a high shelf or in a locked cabinet.

3.3. The Labeling Experiment for Measuring Precautionary Behavior

We designed four different labels for the cleaning agent product and three different labels for the drain opener product. These labels contained different amounts of information about risks and precautions and utilized different formats for presenting the information. By presenting different consumers with different labels and asking each of them about their intentions to take precautions against the risks from the products, we were able to measure the relationship among the information on the labels, the formats of the labels, and the extent of precautionary behavior taken by product users.

Figure 3.1 shows black-and-white reproductions of the four different labels used on the cleaning agent product. All four labels used the same brand name (Vector) and the same logo on the front, and all were colored in red, blue, black, and white. The first label contains no information about the possible hazards of using the cleaning agent and the precautions necessary to avoid them. Any precautions taken by consumers presented with this label would result solely from behavior learned from their prior use of other, similar products.

The second label we call the "Clorox" label because it incor-

WHAT HOUSEHOLD CLEANING JOBS CAN VECTOR DO?

Use Vector to clean your bathroom and kitchen. Vector is an excellent
stain remover and cleaner, yet is economical to use. Vector cleans by
removing stubborn stains and cleans surfaces all around the house.
TOILET BOWLS - Pour in ½ cup of Vector. Brush entire bowl. Let stand
10 minutes, flush.

KITCHEN SINKS - Cover stains with water. Pour ½ cup of Vector directly
into standing water. (Vector will not remove rust or pot marks).
FLOORS, VINYL, CERAMIC TILE, WOODWORK, AND APPLIANCES - Clean with a
solution of 3/4 cup Vector per gallon of sudsy water.
BATHTUBS AND SHOWERS - Clean with a solution of 3/4 cup Vector per gallon
of warm water.

YOUR LAUNDRY NEEDS VECTOR FOR THE BROAD RANGE OF LAUNDRY PROBLEMS
YOU ENCOUNTER.   NO OTHER TYPE OF ADDITIVE USED WITH YOUR DETERGENT CAN
GIVE A CLEANER, BRIGHTER WASH

DIRECTIONS FOR USE

How much should you use?
To get the best cleaning results you should use the proper amount of Vector
in the wash water. The guidelines below should provide excellent cleaning
results with any good soap or detergent. However if you wash extremely
heavy soiled or very large loads, you may want to add slightly more Vector.

* Large top loading automatic - 1½ cups   * Regular top loading automatic - 1 cup
* Front loading automatic - ½ cup   * Heavy soil - increase amount by ¼ cup
* Hand laundry - 2 gallons of sudsy water - 1/8 cup

What fabrics can you launder?
Cotton, linen, synthetics, permanent press and most colored fabrics can be
safely laundered. Test any article to determine if it is laundry safe by applying
one drop of tablespoon Vector mixed with ¼ cup water to a hidden part of the
fabric. Be sure to check all colors, including trim. Let stand one minute, then
blot dry, if there is no color change the article can be safely laundered. Do
not use Vector on silk, wool, mohair, leather, spandex, or non-fast colors.
Repeated use on flame-retardant fabrics made of 100 % cotton may cause loss of
flame retardancy.

How should you add Vector to your washload?
Select the way to add Vector that is easiest for you: in the wash water before
the laundry is put in, or diluted with a quart of water after the washer has
begun agitating. For a washer with automatic dispenser, follow the manufacturer's
instructions.

What special laundry problems can Vector help solve?
Particularly tough laundry problems such as blood, berries, perspiration, or
other stubborn stains may require more than one washing with Vector or soaking
for 5 minutes in a solution of ¼ cup Vector to a gallon of cool sudsy water
prior to washing with Vector.

Figure 3.1 Cleaning agent labels. A: Label with no warning information.

WHAT HOUSEHOLD CLEANING JOBS CAN VECTOR DO?

Use Vector to clean your bathroom and kitchen. Vector is an excellent stain remover and cleaner, yet is economical to use. Vector cleans by removing stubborn stains and cleans surfaces all around the house.
TOILET BOWLS - Pour in ⅓ cup of Vector. Brush entire bowl. Let stand 10 minutes, flush. Do not use Vector with other toilet bowl cleaners. See caution statement.
KITCHEN SINKS - Cover stains with water. Pour ¼ cup of Vector directly onto standing water. Vector will not remove rust or pot marks).
FLOORS,VINYL, CERAMIC TILE, WOODWORK, AND APPLIANCES - Clean with a solution of 3/4 cup Vector per gallon of sudsy water.
BATHTUBS AND SHOWERS - Clean with a solution of 3/4 cup Vector per gallon of warm water.

YOUR LAUNDRY NEEDS VECTOR FOR THE BROAD RANGE OF LAUNDRY PROBLEMS YOU ENCOUNTER. NO OTHER TYPE OF ADDITIVE USED WITH YOUR DETERGENT CAN GIVE A CLEANER, BRIGHTER WASH

DIRECTIONS FOR USE

How much should you use?
To get the best cleaning results you should use the proper amount of Vector in the wash water. The guidelines below should provide excellent cleaning results with any good soap or detergent. However if you wash extremely heavy soiled or very large loads, you may want to add slightly more Vector.

* Large top loading automatic - 1¼ cups   * Regular top loading automatic - 1 cup
* Front loading automatic - ½ cup   * Heavy soil - increase amount by ¼ cup
* Hand laundry - 2 gallons of sudsy water - 1/8 cup

What fabrics can you launder?
Cotton, linen, synthetics, permanent press and most colored fabrics can be safely laundered. Test any article to determine if it is laundry safe by applying one drop undiluted Vector mixed with ¼ cup water to a hidden part of the fabric. Be sure to check all colors, including trim. Let stand one minute, then blot dry. If there is no color change the article can be safely laundered. Do not use Vector on silk, wool, mohair, leather, spandex, or non-fast colors. Repeated use on flame-retardant fabrics made of 100 % cotton may cause loss of flame retardancy.

How should you add Vector to your washload?
Select the way to add Vector that is easiest for you. in the wash water before the laundry is put in, or diluted with a quart of water after the washer has begun agitating. For a washer with automatic dispenser, follow the manufacturer's instructions.

What special laundry problems can Vector help solve?
Particularly tough laundry problems such as blood, berries, perspiration, or other stubborn stains may require more than one washing with Vector or soaking for 5 minutes in a solution of ¼ cup Vector to a gallon of cool sudsy water prior to washing with Vector.

> CAUTION. Vector may be harmful if swallowed or may cause severe eye irritation if splashed in eyes. If swallowed, feed milk. If splashed in eyes, flood with water. Call physician. Skin irritation: if contact with skin, wash off with water. Do not use Vector with ammonia or products containing acids such as toilet bowl cleaners, rust removers or vinegar. To do so will release hazardous gases. Prolonged contact with metal may cause pitting or discoloration. Do not use this bottle for storage of any other liquid but Vector.

Figure 3.1 (cont.) *B*: "Clorox" label.

VECTOR CLEANING AGENT

WARNING: NOT FOR PERSONAL USE. DO NOT GET ON SKIN OR IN EYES. DO NOT TAKE INTERNALLY. IF SPLASHED IN EYES OR ON SKIN, FLOOD WITH WARM WATER 10-15 MINUTES. IF IRRITATION PERSISTS CALL PHYSICIAN. IF SWALLOWED, GIVE MILK OR BREAD SOAKED IN MILK FOLLOWED BY COOKING OIL. CALL PHYSICIAN. KEEP OUT OF REACH OF CHILDREN.

IMPORTANT: Do not use cleaning agent in combination with ammonia or other household chemicals, toilet bowl cleaners, rust removers, etc., since if such mixing occurs, hazardous gases may be released. Avoid transfer to food or beverage containers. Avoid contamination of food. KEEP IN A COOL PLACE.

TEST FOR LAUNDERABILITY: To test a colored fabric, wash and wear cottons and rayons, or a fabric of unknown composition for launderability, apply a mixture of 1 tablespoon of Vector in a gallon of hot water to an inconspicuous corner and let stand for 3-5 minutes. If fabric color fades or yellows, it is not Vector-safe.

PURPOSE	AMOUNT	DIRECTIONS
LAUNDERING: To launder white and colorfast cotton, linen, nylon, Dacron, Orlon and rayon in washing machine.	1 cup VECTOR per load for conventional washer. ½ cup for front load automatic.	Add to pre-soak, wash water or first rinse. If clothes are in machine, dilute in 1 qt. water, then add.
TO WHITEN "AGE YELLOWED" NYLON:	1 tablespoon Vector per gallon water.	Soak clean fabric in solution 15-20 minutes. Rinse well. Repeat if necessary.
TO REMOVE STAINS: Fruit berry, wine, coffee, tea, ink, grass, dye, medicine stains, scorch, etc.	Make a solution of 2 tablespoons VECTOR to each quart of water.	Immerse fabric for 5-10 minutes. Rinse well. Repeat if necessary.
TO CLEAN KITCHEN AND BATHROOM: Refrigerator, tile, and tub.	2 tablespoons VECTOR to 1 quart water.	Wash, rinse and dry. Do not use on silverware.

DISCARD EMPTY CONTAINER AFTER THOROUGHLY RINSING WITH WATER.

DO NOT USE ON SILK, WOOL, ACETATE, SPANDEX OR LEATHER.

Distributed by The Steuart Co.
Pittsburgh, PA 15230

Figure 3.1 (cont.) C: "Bright" label.

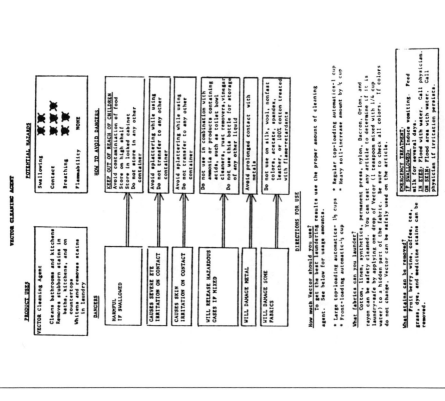

Figure 3.1 (cont.) *D*: Test label.

porates a slightly modified version of the risk information on Clorox brand bleach for our Vector brand cleaning agent. Clorox is the best-selling brand of bleach in this country.

The third label we call the "Bright" label because it is a modified version of the label on the Kroger Company's Bright brand of bleach. Since the format of the information on the Bright label is quite different from the format of similar information presented on the Clorox label, comparing the precautions taken by consumers receiving these two types of labels would enable us to make a direct test of the importance of formatting in label design.

The fourth label, which we call the "Test" label, presents the same information as the Clorox and Bright labels but uses a format that the principles of consumer decision theory indicate should be more effective. The Test label organizes all the usage information systematically and explicitly links the risks and appropriate precautions. We structured the new format to improve the label's effectiveness regarding all product uses, not just those that were safety related.

Figure 3.2 shows copies of the three different labels attached to the drain opener product. All of them use the same brand name, Unstop, and the same logo on the front panel. The first label contains only product content and usage information; it gives no information about the hazards of incorrect use or the precautions necessary to avoid those hazards. The second label we call the "Drāno" label because it incorporates the risk and usage information on Liquid Drāno brand drain opener. Because we formulated our drain opener as a more powerful chemical than Liquid Drāno, actually closer to a liquid form of lye, the Drāno label is a combination of the information contained on the Liquid Drāno package and the crystalline Red Devil brand lye package. The final Test label for drain opener was also designed in accordance with the principles of effective label design. It differs from the Drāno label only in format.

3.4 Sample Description and Interview Method

To collect data on the differences in precautionary responses to the seven different labels on the two products, we interviewed 368 subjects recruited by a consumer research firm at a mall in Greensboro, North Carolina. The mall drew shoppers from a broad cross-section of demographic groups in the Greensboro area. The Greens-

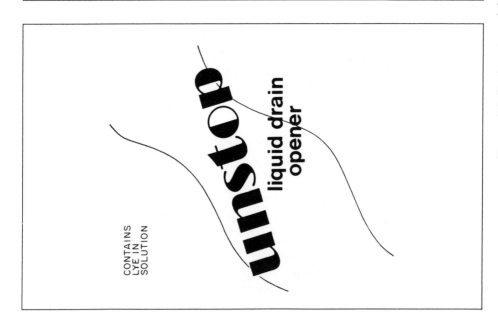

DIRECTIONS FOR USE

TO OPEN: Press down on cap while turning in a counter-clockwise direction.
TO LOCK: Turn cap onto threads in a clockwise direction until it no longer turns. Then press down on cap and turn to seal.

KEEP DRAINS OPEN: EVERY WEEK remove drain sieve. Pour ½ bottle into drain. Let stand 10 minutes. Flush with hot water from faucet.

OPENING CLOGGED DRAINS: Remove drain sieve and any standing water from sink, washbowl, or tub. Use ½ bottle. Allow to work 30 minutes. Flush with hot water when drain clears. Repeat if necessary.

Contents: Sodium hydroxide and water

44600 00102

CONTAINS
LYE IN
SOLUTION

unstop
liquid drain
opener

Figure 3.2 Drain opener labels. A: Label with no warning information.

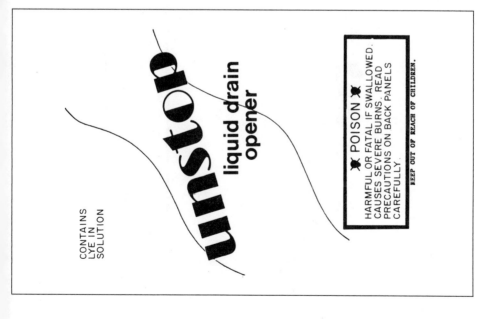

Figure 3.2 (cont.) *B*: "Drāno" label.

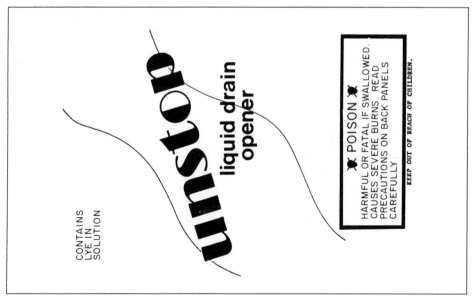

Figure 3.2 (cont.) C: Test label.

boro area is often used as a test site for national marketing studies because of its representative character.

All participants in the survey were screened to ensure that they used at least some household cleaning products. Our interviewers brought the subjects into an enclosed room within the mall, handed each of them one product (either a cleaning agent or a drain opener), and proceeded through one of the questionnaires reproduced in Appendixes A and B. In stage 1 of the questionnaire each subject was asked to examine the product as if he or she was about to use it at home for the first time. Much of the questionnaire was administered by an interviewer from a marketing firm, but for the questions about the price-risk tradeoff we used an interactive computer program. This approach (discussed below) eliminated potential interviewer bias and enabled us to have the computer in effect redesign the questionnaire to elicit the desired information about tradeoffs.

Measuring Precautionary Behavior

Stages 2 and 3 of both questionnaires contain the questions that were asked by the interviewer to determine which precautions, if any, the subject would take in using the particular product. For the cleaning agent, we were interested in determining whether the subject would use the product with a toilet bowl cleaner or some other ammonia-based cleaner (questions 2.4 and 2.6). We included the other four questions in stage 2 to disguise the intent of our questions concerning the precautions against mixing. Similarly, stage 2 of the drain opener questionnaire was primarily concerned with determining whether the subject would wear rubber gloves when using the drain opener. We added questions 2.1, 2.2, and 2.3 to the question of interest (2.4) in order to disguise the main intent of this set of questions.

In stage 3 of both questionnaires we asked the open-ended question: "Where would you store a product like this in your home?" The interviewers probed enough to determine whether or not the subject would store the product in a childproof location, without cuing the subject about the information we were seeking.

To assess whether the consumer responses to the various product labels were influenced by the characteristics of the subjects participating in our survey, we used the demographic data from the questionnaire to form several explanatory variables in the equations reported in the next two chapters. Of the 368 subjects recruited for the mall interview, 200 were given the cleaning agent

Table 3.2 Sample characteristics

Variable	(1) Sample mean	(2) Standard Deviation	(3) Population mean
Family income[a] (1984 $/year)	30,828	17,932	28,557
Education[a] (years)	13.3	2.2	12.5
Age[a] (years)	33.3	12.8	30.0
Married[a] (1 = yes, 0 = no)	0.57	0.50	0.64
Male[a] (1 = yes, 0 = no)	0.32	0.43	0.47
Black[a] (1 = yes, 0 = no)	0.26	0.40	0.12
Number of children under 5 per family[a]	0.23	0.51	0.29
Bleach use per family[b] (bottles/year)	17.0	31.3	8.1
Drain opener use per family[b] (bottles/year)	2.0	4.4	2.4

a. Source of U.S. population statistics: U.S. Bureau of the Census, *Statistical Abstract of the United States* (Washington, D.C.: U.S. Department of Commerce, 1985) (1983 data, with family income adjusted to 1984 $s).

b. Source of U.S. population statistics: *Predicasts Basebook* (Predicasts: Cleveland, 1984).

questionnaire and 168 were given the drain opener questionnaire. Table 3.2 lists the characteristics of the overall sample.

The means for the subjects receiving each of the two products (not shown) do not differ significantly from the overall means shown in the table. Although we do not have data on the demographic and usage characteristics of the population of bleach and drain opener users in the entire country, the statistics in Table 3.2 appear to be reasonably representative of the total population of product users. The sample means in column 1 match the U.S. population means in most dimensions, and the differences are consistent with the likely differences between the U.S. population and the population of product users. For example, bleach and drain opener tend to be used more by women than by men, and because some families do not use bleach, the number of bottles used per consuming family exceeds the total use divided by the number of U.S. families.

Measuring Willingness to Pay to Avoid Taking Precautions

If there were no time, effort or other source of disutility associated with taking the precautions necessary to avoid injuries from hazardous products, consumers who were aware of precautions would always take them. However, most consumers do attach some disutility to taking many precautions, although the levels of aversion differ across consumers and across the precautions themselves. A rational consumer will take only precautions with costs less than their benefits. Assessing the tradeoff requires a comparison of the two components of the decisionmaking problem.

Stage 4 of both questionnaires was designed to measure each subject's willingness to pay to forgo taking the precautions necessary for avoiding the hazards. We asked each subject to consider two products, called current and new, and identical in all respects except one, and then to answer how much he or she would pay (over the price of the current product) to obtain the new product with the single more desirable characteristic. This approach, called contingent valuation, has been widely used by environmental economists to assess consumer preferences directly. (For further discussion of the contingent valuation approach, see the next section and Chapter 5.)

Again, in order to disguise the intended use of the answers elicited in this section, we added three extraneous questions (4.1, 4.2, and 4.3) to stage 4 on both questionnaires. Our intent was to force the subjects to focus on one characteristic of the product at a time, which would make them less likely to mix their valuations of its convenience benefits (specifically, avoiding having to take the four precautions) with any inferences they might make from this characteristic about the relative risk levels of the current and new products.

Stage 5 asks questions about the frequency of product use.

Measuring Willingness to Pay to Avoid Accidents

We used two approaches in stage 6 to measure consumer willingness to pay to avoid the injuries associated with hazardous chemical products—contingent valuation and conjoint analysis of paired comparison data (Randall, Hoehn, and Brookshire 1983; Green and Srinivasan 1978). In our application, the two approaches differ in that contingent valuation poses the valuation question directly, though in a contingent market environment. For example, for a new product differing from a current one only in its associated injury levels and prices, we asked how high the price of the new product would have to be before the subject would rather buy the current one. This approach, which has been used frequently by economists in contexts not involving risks, approaches the price-risk tradeoff directly.

In contrast, the conjoint approach gives the consumer information about all relevant characteristics of the alternative products available and asks him or her to rank them by order of overall attractiveness. We asked subjects to rate several pairs of products on a scale; the characteristics of the products in each paired comparison were designed to allow us to recover the preferences for characteristics from each consumer's ranking. This conjoint approach is one of the dominant techniques used in marketing to elicit consumer attitudes because it replicates the kinds of tasks that consumers undertake when making actual economic decisions.

Each subject was asked questions about the two hazards associated with the particular product received. Subjects given the cleaning agent were asked questions that allowed us to value how much they were willing to pay to avoid both chloramine gas poisoning and child poisonings. Subjects receiving the drain opener were asked about their willingess to pay to avoid hand burns and child poisonings.

Within each group of subjects receiving the same product, about half were asked the contingent valuation (CV) question first and then the conjoint questions, and half were asked the conjoint questions first. This design allowed us to analyze whether answering the conjoint questions first biased the CV responses, and, vice versa (these issues are addressed more fully in Magat, Viscusi, and Huber 1985). We always asked the questions about the child poisoning injury after asking evaluation questions about the chloramine gas poisoning (for the cleaning agent) or the hand burns (for the drain opener).[3]

3.5 Summary

In our consumer information study we surveyed consumers of household cleaning products to measure their responses to the information on two different products, a liquid cleaning agent and a liquid drain opener, which we judged to be most useful in learning about how consumers respond to hazard labeling. Two primary hazards from misuse were identified for each product, along with the precautions necessary to avoid these hazards. Four cleaning agent labels and three drain opener labels were used to test consumer responses to variations in information content and format on the labels. In the main parts of the survey we asked questions measuring precautionary behavior in response to the product labels, willingness to pay to avoid taking precautions, and willingness to pay to avoid accidents. The results from this study form the basis for the analysis presented in the next two chapters about how risk information on the labels of hazardous chemical products affects consumers' precautionary behavior and about how consumers trade off lower product prices and improved product safety.

4 | The Effect of Risk Information on Precautionary Behavior

W. KIP VISCUSI,
WESLEY A. MAGAT,
and
JOEL HUBER

This chapter uses the responses to the survey described in Chapter 3 to analyze how the content and format of the information on product labels affect the precautionary behavior of consumers who use them. Provision of risk information will have diverse effects. Risk perceptions may change, affecting a range of individual decisions. Consumers may alter the products they purchase, and workers may switch jobs. Even without such a drastic change in activities, there may be important behavioral effects. The effect of greatest concern for pesticide and chemical regulation, for example, is how risk information influences the safety precautions that individuals will select. In this chapter we address this effect on precautionary behavior; subsequent chapters focus on consumer decisions to purchase a product and workers' job choices.

The principal appeal of informational policies is that they do not in themselves constrain the choices that are available. To the extent that inadequate information is the source of the market failure, the hazard warning regulation addresses the source of the market failure directly without disturbing the other beneficial features of the market. As a result, users of a hazardous product with

different susceptibilities to a particular hazard, different preferences toward risk, and different product needs and usage rates can
select the combination of risk, product efficacy, and usage rates
that is best for each of them. The informational strategy is often
viewed as a stopgap measure until the health implications of the
hazard are better understood, thus permitting greater flexibility
in situations in which our knowledge of risks is evolving over time.
In situations in which our knowledge about a risk is not sufficient
to warrant a ban but does provide cause for concern, policymakers
often adopt labeling as an intermediate course.

For the hazard information effort to be successful, however, individuals must be able to think systematically about risks and to
make sound decisions under uncertainty. As discussed in Chapters
1 and 2, an increasing number of studies indicate that decisions
under uncertainty are notoriously difficult.

Yet little evidence exists on the behavioral implications of in-
formation-oriented policies. Many informational campaigns have
been education efforts that have provided little new knowledge
and have yielded disappointing results.[1] The research results reported here are pertinent both to the debate over the degree to
which safety precautions are governed by economic factors and to
the more general issue of whether individuals make rational decisions under uncertainty.

In this chapter we explore the nature of this behavioral linkage
between risk information and safety precautions. We consider not
only this fundamental relationship but also related hypotheses,
such as how the amount of risk information provided influences
behavior. After presenting the underlying theoretical model in Section 4.1, we discuss the effects of the labels in Sections 4.2 and 4.3.
Many of the key findings in the chapter are presented in the tables
in Section 4.2. Possible shortcomings in consumers' information
processing are explored in Section 4.4. The results, which are broadly
consistent with a Bayesian framework, are summarized in Section
4.5.

4.1 A Conceptual Model of the Effect of Information

The analysis of individual responses to hazard warnings is similar
in spirit to economic models of seatbelt use and the effects of safety
caps on consumer precautions.[2] In each case, either the change in
the individual's perceived safety or an actual shift in safety, which

in turn influences perceptions, leads individuals to choose appropriate degrees of care. Increases in the perceived risk associated with an activity will lead to greater incentives to exercise care as the benefits of undertaking costly precautionary behavior have risen.

Here we develop a simple model to analyze how the implied risk and informational content of the hazard warning affect individual behavior. Let e be the level of precautionary effort, which imposes a disutility $V(e)$ on the consumer, where $V', V'' > 0$. Precautions have a beneficial effect insofar as they reduce the probability of an accident. Let $q(e)$ be the consumer's assessed probability of an accident before receiving the hazard warning information, where $q' < 0$ and $q'' > 0$. The consumer's income is I, and the size of the loss associated with an accident is L, so that the payoff for a risk-neutral consumer is $I - V - L$ if there is an accident and $I - V$ if there is not.

The hazard warning influences a consumer's precautionary behavior through its impact on his risk perceptions. Assuming that individuals' probabilistic beliefs can be characterized by a beta distribution, which is ideally suited to stochastic processes such as this, let γ represent the precision of the consumer's prior probabilistic beliefs concerning the risk of the product. In effect, the consumer acts as if he has drawn γ balls from an urn, where a fraction $q(e)$ of these represent his subjective assessment of the probability of an accident. The hazard warning information also has an associated risk and informational content. Let $sq(e)$ be the risk of the accident implied by the warning, where s is a relative risk scale factor and

$$0 \leq sq(e) \leq 1.$$

Values of s above 1 indicate warnings that have an implied risk of the accident above the consumer's prior assessed risk.

The informational content of the hazard warning is ξ, so that the consumer's assessed risk of the product after reading the warning label is

(4.1) $$p(e) = \frac{\gamma q(e) + \xi s q(e)}{\gamma + \xi} = \frac{q(e)(\gamma + \xi s)}{\gamma + \xi}.$$

It is instructive to rewrite this expression, letting Ψ equal the informational content of the warning relative to the consumer's prior beliefs (that is, $\Psi = \xi/\gamma$), so that

(4.2)
$$p(e) = \frac{q(e)(1 + \psi s)}{1 + \psi}.$$

The consumer's effort decision is to maximize his net rewards Z, or

$$\text{Max } Z = I - V(e) - p(e)L,$$
$$e$$

yielding the requirement

(4.3)
$$V_e = -p_e L,$$

or the marginal disutility of effort equals the incremental value of the safety improvements. As expected, the optimal degree of pre-caution-taking increases with the size of the loss, or

(4.4)
$$\frac{\partial e}{\partial L} = \frac{-p_e}{p_{ee}L + V_{ee}} > 0.$$

In addition, increasing the disutility of effort associated with precautions reduces the optimal effort choice. Letting b be a shift parameter that raises $V(e)$ to $bV(e)$, the effect of b is to lower the effort choice, or

(4.5)
$$\frac{\partial e}{\partial b} = \frac{-V_e}{p_{ee}L + bV_{ee}} < 0.$$

The matters of primary interest are the effects of the two warning parameters, s and Ψ, on effort, which are given by

(4.6)
$$\frac{\partial e}{\partial s} = \frac{-p_{es}L}{p_{ee}L + V_{ee}}$$

and

(4.7)
$$\frac{\partial e}{\partial \psi} = \frac{-p_{e\psi}L}{p_{ee}L + V_{ee}}.$$

Three conclusions follow from equations (4.6) and (4.7). First, the responsiveness of precautionary effort to changes in both pa-rameters of the warning increases with the size of the loss, that is:

(4.8)
$$\frac{\partial}{\partial L} \left| \frac{\partial e}{\partial s} \right| > 0$$

and

(4.9)
$$-\frac{\partial}{\partial L}\left|\frac{\partial e}{\partial \psi}\right| > 0.$$

Second, boosting the implied risk s associated with the warning relative to the consumer's own prior estimates raises his level of safety effort because

(4.10)
$$p_{es} = \frac{\psi}{1 + \psi} q' < 0.$$

Finally, the effect of stronger informational content Ψ of the warning depends on the associated implied relative risk because

(4.11)
$$p_{e\psi} = \frac{(s - 1)}{(1 + \psi)^2} q'.$$

If the implied relative risk is above 1, more informational content of the message increases precautionary effort; if it is below 1, precautionary effort decreases. This result is explained by the fact that with an above-average relative risk ($s > 1$), more information raises the consumer's posterior (that is, subsequent) estimate of the product's risk (see Eq. 4.1), whereas a more informative message lowers the consumer's risk perception when prior estimates are above average ($s < 1$). In the empirical analysis below, we address the influence of both the risk level and the informational content on precautionary behavior.

4.2 The Effects of Labels on Precautionary Behavior

Overview of the Experiment

Both household bleach and drain opener pose substantial short-term risks. The major risks associated with household bleach are twofold. First, individuals who drink bleach (typically children under the age of five) typically vomit and experience stomachaches for about a day. This injury can be prevented by keeping the products out of the reach of children, for example, on a high shelf. Chloramine gas poisonings pose the second risk. If bleach is mixed with ammonia or ammonia-based products, chloramine gas forms, leading to headaches, burning lungs and eyes, and possibly hospitalization for several days. This risk can be avoided by not mixing bleach with toilet bowl cleaners or other ammonia-based products.

The two main risks of the drain opener also affect both children and the product users. A child poisoning from drain opener leads

Table 4.1 Summary of products and label characteristics

Label	Percentage of risk information (WARNAREA)	Label format
Bleach		
No warning	0	standard
Clorox	31	standard
Bright	41	standard, but more prominent risk information
Test	69	formatted to highlight uses, risks, and precautions
Drain opener		
No warning	0	standard
Drāno	78	standard
Test	63	formatted to highlight uses, risks, and precautions

to severe and painful burns to the mouth and throat, possibly including loss of the esophagus. This outcome was the most severe health effect analyzed in the consumer portion of the study. Spilling the product on one's hand produces painful burns and red swollen blisters that heal in about a week. The recommended precaution is to wear rubber gloves.

In addition to one label lacking any risk information, we included labels with different formats apprising consumers of the pertinent hazards associated with the product. An important difference among the labels from the standpoint of this chapter is the variation in the amount of informational content. As the summary in Table 4.1 indicates, the fraction of the label devoted to risk information varied considerably, reaching as high as 78 percent for a label modeled after the existing labels for Drāno and Red Devil Lye (hereafter called the Drāno label). Most of these differences pertained to the size of the type used and the degree of repetition of the warning throughout the label. The last column in Table 4.1 summarizes the differences in format, which is also a major concern. The bleach labels patterned after Clorox brand bleach and the Kroger grocery chain's house brand, Bright Bleach, were of standard format, but the Bright label located the risk information more prominently. The Drāno label likewise gave the risk information prominence.

Table 4.2 Effects of labels on precaution-taking: Bleach (percentages)

Precaution	No warning (n = 51)	Clorox (n = 59)	Bright (n = 42)	Test (n = 44)	Maximum incremental effect
1. Do not mix with toilet bowl cleaner (if toilet is badly stained)	16	23	36	40	24
2. Do not add to ammonia-based cleaners (for particularly dirty jobs)	69	68	69	84	16
3. Store in childproof location	43	63	50	76	33

Mean Effects

Table 4.2 summarizes the effects of the four bleach labels on the key precautions. The precautions necessary to avoid chloramine gassing were measured through two separate questions (numbers 1 and 2 in the table). In addition, the chloramine gas precautions were related to conditional behavior for unusual circumstances (badly stained toilets and particularly dirty jobs), so that these results are conditional on particular cleaning situations. The extent of misuse in practice will be below the fraction of consumers who indicate potential misuse in the questionnaire, because these contingencies may not arise. The four label formats were those involving no hazard warnings, the Clorox label, the Bright label, and the Test label.

The responses in Table 4.2 measure consumer intentions to take precautions, rather than the fraction who actually take the precautions. Ideally, we would have liked to ascertain the final behavioral response. However, past studies and evidence discussed in Chapter 6 suggest that in carefully designed surveys closely linked to actual decisions, the hypothetical responses parallel actual behavior.[3] Moreover, in a subseqent study in which we compared the survey results with estimates of precautionary behavior elicited through a telephone survey about products existing in the marketplace, we found an extremely close correspondence between the behavioral intentions for the experimental product labels that wcre patterned after existing products on the markets and the telephone estimates of actual behavior (see Viscusi and Magat 1986). Finally, the nature of the research issues, including analysis of labels lacking hazard warnings, necessitated such an experimental approach.

Although the experimental design restricted the role of prior consumer knowledge by calling the product a cleaning agent rather than a bleach, it is clear that consumer familiarity with similar cleaning products had some influence on behavior. Even in the group given no hazard warning, 16 percent of subjects said they would not mix the cleaning agent with toilet bowl cleaner, 69 percent said they would not mix it with ammonia-based cleaners, and 43 percent said they would store it in a childproof location. The responses to the questions about mixing the product with toilet bowl cleaner and ammonia-based cleaner may include consumers who did not envision the need for ever mixing the product in that fashion rather than a reluctance stemming from safety-related reasons. In contrast, the storage in a childproof location response presumably would reflect this prior knowledge of the risk to a greater extent. These results are supportive of a hypothesis that consumers learn from labels, because they suggest that there is consumer awareness of differences in risk across product classes. Insofar as existing labels have contributed to this knowledge base, our results understate the incremental effect of labels in situations in which consumers have never read similar labels.

Of all the bleach labels, the Test label performed best in terms of inducing consumers to take precautions. The Clorox and Bright labels had modest effects on the chloramine gas risks from mixing bleach with toilet bowl cleaner, and the Test label more than doubled the fraction of subjects who would undertake this precaution.[4] Nevertheless, more than half of the subjects said they would not take the precaution despite the warning on the Test label. This last result does not imply that with the Test label 60 percent of the consumers would misuse the product. The original question was conditional, dealing with use of the cleaning agents for "badly stained" toilets. If this contingency did not arise, the potential misuse might not occur either.

Consumers appeared to be much less likely to mix the cleaning agent with ammonia-based cleaners other than toilet bowl cleaners. However, the two labels that are now used to alert consumers to the chloramine gas dangers from such a mixture appear to have had no more beneficial effect than the label with no warning. Only

Table 4.3 Effects of labels on precaution-taking: Drain opener
(percentages)

Precaution	No warning (n = 59)	Drāno (n = 59)	Test (n = 50)	Maximum incremental effect
1. Wear rubber gloves	63	82	73	19
2. Store in childproof location	54	68	66	12

the Test label showed any impact. It increased the fraction of subjects who would not add the product to ammonia-based cleaners by a statistically significant 16 percent, bringing to over four-fifths the fraction of subjects who would not undertake such a mixture.

All three labels with risk information increased the percentage of subjects who planned to store the bleach in a childproof location, although the Bright label was not significant at traditional confidence levels. The Test label was most effective; it increased the childproof storage propensity by 33 percent over that with the label containing no warnings. The Test label created an awareness of the key risks among over one-tenth of the subjects who would not otherwise have been reached with existing labels. To the extent that consumer responses to the new information format on the Test label would be enhanced by longer-term familiarity with its new formatting, these results understate the label's eventual effectiveness.

The necessary precautions for avoiding the risks of drain openers appear to be well known to consumers even in the absence of a warning on the label. Table 4.3 shows that the majority of subjects (63 percent) would wear gloves even in the absence of the warning and would also store the product in a childproof location (54 percent). The higher precautionary storage response for drain opener than for bleach in the absence of a hazard warning suggests that consumers make no simple uniform response to childproofing warnings independently of the product. Rather, prior knowledge of the greater severity of the health impact of child poisonings from drain opener than from bleach accounts for the difference. This result accords with our conceptual analysis, since the optimal safety effort increases with the size of the potential loss.

As with bleach, with drain openers there was evidence of differential performance of the labels that contained risk information. The Drāno label increased the propensity to wear rubber gloves by 19 percentage points and to store the product in a childproof location by 12 points. In contrast, the Test label had roughly half this increase and differed from the label with no risk warning by statistically insignificant amounts.

Clearly, labels have a consistent differential impact on precautionary behavior. For both products examined, labels including risk information generally led to an increase in safety precautions, whereas labels lacking risk warnings did not. (The differences in the effects of the various labels are discussed further below.) In addition, labels did not lead all consumers to take precautions, a result that is also to be expected if there is heterogeneity in the desirability of precautions.

4.3 Variations in Precautionary Behavior

Probability Equations

Because of the randomized nature of our sample, the results in Tables 4.2 and 4.3 provide a fairly accurate measure of the incremental effects of the labels. Randomization serves to distribute individuals with different attributes across the labeling treatments so that with perfect randomization and a sufficiently large sample it should not be necessary to perform a multivariate analysis to distinguish the effect of labels. Such an analysis is instructive, however, to identify whether any variables other than labeling format influenced the propensity to take precautions. Moreover, to the extent that the limited sample was not large enough to ensure sufficient randomization, this procedure better isolates the incremental effect of a particular label.

From the results in Section 4.1, we know that variables associated with a lower disutility of effort, greater losses from an accident, or higher relative risk of an accident should lead to increased precautionary behavior. We examined the relationship between precaution-taking and seven demographic variables—the respondent's age (*AGE*), sex (*MALE* = 1 if respondent is male), race (*BLACK* = 1 if respondent is black), marital status (*MARRY* = 1 if respondent is married), years of schooling (*EDUC*), family income (*INCOME*), and the number of children in the high-risk poisoning group—children under five years of age (*FIVE*).

On average, female users are probably more likely to take precautions, such as wearing gloves, because to the extent that they wear rubber gloves regularly, the inconvenience or disutility costs of doing so for drain cleaner are less. The influence of household wealth is captured by several variables: *BLACK, AGE, INCOME,* and *EDUC*. Because of the positive income elasticity of demand for health, the magnitude of any health loss is greater for wealthier consumers. As a result, from Section 4.1 we know that more affluent consumers are consequently more likely to take precautions. However, the disutility cost associated with precautions may be greater for richer consumers, particularly if the precautions involve time allocations, which will impose a greater opportunity cost on this group. The net effect of wealth is thus unclear theoretically.

Perhaps the variables whose influences are most clear-cut are *MARRY* and *FIVE*. Households with children under the age of five and possibly households where the product user is married will be more likely to have children exposed to the hazardous products (that is, the household's relative risk of an accident is higher) and consequently will have a greater incentive to take care.

The labeling impacts were captured with three label dummy variables (*CLOROX, BRIGHT,* and *TEST*), whose coefficients are given relative to the no warning label. We also explored an interactive effect of the labels with *EDUC* to ascertain whether only better-educated consumers were influenced by the risk information, but these results are not reported, because no significant influences were observed.

Because of the discrete nature of the dependent variable pertaining to whether or not the respondent would take particular types of precautions, we utilized a logit estimation procedure to estimate the probability of taking precautions. Table 4.4 presents the maximum likelihood estimates for each of the risk-related actions for bleach. It shows that the label effects follow the same general pattern as the mean precaution-taking percentages in Table 4.2. The propensity to avoid mixing bleach with toilet bowl cleaner is increased most by the Test label, with the Bright label ranking next in effectiveness. The Clorox coefficient is not statistically significant at the usual levels and is much smaller in magnitude. The only label with a substantial positive effect on the propensity to add bleach to ammonia-based cleaners is the Test label, but the asymptotic standard error on this coefficient is quite large. The labeling results are strongest for storage in a childproof location. The Test label boosts this precaution dramatically, and the Clorox label substantially. Overall, the close parallels between the mul-

Table 4.4 Maximum likelihood estimates of precaution probability equations for bleach

Independent variable	Coefficients (asymptotic std. errors)		
	Do not mix with toilet bowl cleaner	Do not add to ammonia-based cleaners	Store in childproof location
Intercept	−2.468	−0.901	−0.709
	(1.271)	(1.240)	(1.155)
AGE	−0.0005	0.012	0.012
	(0.015)	(0.015)	(0.013)
MALE	−0.410	−0.104	0.554
	(0.418)	(0.391)	(0.377)
BLACK	−1.252	−0.521	0.087
	(0.545)	(0.426)	(0.422)
MARRY	−0.248	0.560	−0.328
	(0.369)	(0.362)	(0.352)
EDUC	0.112	0.141	−0.042
	(0.086)	(0.087)	(0.079)
INCOME	-6.2×10^{-6}	2.2×10^{-5}	4.8×10^{-6}
	(13.6×10^{-6})	(1.2×10^{-5})	(11.8×10^{-6})
FIVE	—	—	1.707
	—	—	(0.480)
CLOROX	0.686	0.020	0.975
	(0.530)	(0.431)	(0.417)
BRIGHT	1.151	−0.104	0.452
	(0.543)	(0.482)	(0.463)
TEST	1.446	0.809	1.556
	(0.530)	(0.529)	(0.477)
−2 log likelihood	211.8	219.5	237.2

tivariate results in Table 4.4 and the mean precaution frequencies in Table 4.2 suggest that the randomization procedure was an effective means for distinguishing main effects.

The personal characteristic variable of greatest interest—FIVE— should increase the likelihood of storing the product in a childproof location, because households with children in this age group are a higher-risk group. As expected, this variable has a significant positive effect in Table 4.4. Indeed, the presence of a child under the age of five has a greater incremental effect on precaution-taking than does the Test warning label.

The other personal characteristic variables were not particularly influential and, to the extent that there were significant effects,

Table 4.5 Maximum likelihood estimates of precaution probability equations for drain opener

Independent variable	Coefficients (asymptotic std. errors)	
	Wear rubber gloves	Store in childproof location
Intercept	−0.137	−1.467
	(1.285)	(1.211)
AGE	0.014	0.014
	(0.016)	(0.014)
MALE	−0.882	—
	(0.428)	—
BLACK	−0.995	−0.220
	(0.434)	(0.409)
MARRY	0.222	0.502
	(0.402)	(0.390)
EDUC	−0.012	0.045
	(0.086)	(0.080)
INCOME	1.6×10^{-5}	1.7×10^{-6}
	(1.5×10^{-5})	(12.2×10^{-6})
FIVE	—	1.060
	—	(0.473)
DRĀNO	1.558	0.837
	(0.470)	(0.417)
TEST	0.948	0.659
	(0.453)	(0.435)
−2 log likelihood	176.4	199.9

arose only in isolated cases. For example, the only significant impact of household income is one negative coefficient, which runs counter to our expectation of a positive income elasticity of demand for health. Higher-income consumers were more likely to add bleach to ammonia-based cleaners, possibly because they are more likely to own a variety of cleaning products. Black consumers were more likely to mix bleach with toilet bowl cleaners, but no other coefficients were significant (at the 5 percent level).

The multivariate logit equations for the precautions with drain opener, reported in Table 4.5, reveal similar patterns of influences. The labeling treatments were coded as dummy variables, as before. Once again, the labeling format exerted the dominant influence, with the Drāno label being more effective than the Test label. The greatest relative difference between these two formats is for the precaution to wear gloves; childproofing differences are much narrower, less than half of the estimated standard errors of the coefficients.

As in the case of the bleach results, most of the demographic variables were not consequential in Table 4.5. No income effects were observed in either direction, and the only demographic influences other than *FIVE* that were statistically significant for drain opener usage were that males and blacks were less likely to wear gloves. This first influence may be a consequence of the fact that women wear rubber gloves more often than men do. As a result, the inconvenience and any associated discomfort of wearing gloves may be less because they are used to wearing gloves for many household chores. It is noteworthy, for example, that the marketing efforts for rubber gloves are directed primarily at white females.

The background variable of greatest interest is *FIVE*. Once again, respondents with children under the age of five are more likely to store the product in a childproof location. Since this group is the high-risk population for poisonings, the higher responsiveness of this group to the product risk accords with behavior predicted by our model in Section 4.1.

Table 4.6 summarizes the degree to which the labels with hazard warnings succeed in generating a precautionary response among such consumers. Without any hazard warning on the label, over two-thirds of all parents with children under age five would take precautions when using both products, compared with under half for those without children in this age bracket. Among the subjects given the hazard warning on the label, the average rate of precau-

Table 4.6 Family composition and childproofing precautions (percentages)

Label format	Households with children under 5	Households with no children under 5
Bleach		
No warning	67	37
Bright	100	40
Clorox	91	57
Test	92	70
Drain opener		
No warning	70	48
Drāno	90	63
Test	83	61

tion-taking across the five products with warnings was 91 percent.

This substantial effectiveness of the hazard warning on the targeted population accords with the predicted behavior, but the failure of the drain opener warning to produce greater precautionary behavior than the bleach warning does not. Drain opener poses more severe losses as a result of child poisoning than does bleach. Thus, if the disutility of precautions is similar, from equation 4.4 the degree of precaution-taking should be greater for the drain opener than for the bleach. Also, from equation 4.9, if the disutility of the two precautions is similar but the drain opener causes greater losses, then the incremental effect of the hazard label for any given initial level of precautions should be greater for the drain opener.

There are two possible explanations for the greater effectiveness of the bleach labels. The first explanation is based on the informational content of the two sets of labels with respect to the child-proofing precaution. The risk to children was featured prominently on the bleach label, including a section on the front, whereas the drain opener label listed child poisoning as one of many hazards. The amount of hazard warning information specifically related to child poisonings was greater and was featured more prominently on the bleach labels. For both products it is likely that consumers underestimated the risks before reading the labels, so $s > 1$. Thus, from equations 4.7 and 4.11 it follows that the bleach label's greater informational content for child poisonings could have outweighed the loss effect and led to a somewhat larger effect than the drain opener label, where the implied risk (that is, loss from the accident) was greater but the informational content was less.

The second explanation is based on the different amount of learning about the dangers of child poisoning that the two sets of labels induced. If consumers initially had a higher assessment of the hazards to children from drain opener than from bleach, then from equations 4.6 and 4.10 the bleach labels would have caused a larger revision in their prior estimates (that is, a higher value of s), which would have led to a greater increase in precautionary effort if this learning effect dominated the loss effect.

The Role of Informational Content
The importance of the amount of the label's informational content is further supported by an econometric analysis that replaces the earlier label dummy variables with a variable, *WARNAREA*, that captures the amount of risk information provided. Specifically, we defined *WARNAREA* as the share of the label area devoted to hazard

Table 4.7 WARNAREA effects on precaution-taking[a]

Precaution	WARNAREA coefficients (asymptotic std. errors)	SPECWARN coefficients (asymptotic std. errors)
Bleach		
Do not mix with toilet bowl cleaner	0.021 (0.007)	0.0695 (0.254)
Do not add to ammonia-based cleaners	0.010 (0.007)	0.053 (0.031)
Store in childproof location	0.020 (0.006)	0.056 (0.020)
Drain opener		
Wear rubber gloves	0.018 (0.005)	0.017 (0.015)
Store in childproof location	0.011 (0.005)	0.034 (0.016)

a. Other variables included in the equation were the same as in Tables 4.4. and 4.5, except that the labeling dummy variables were excluded. The WARNAREA and SPECWARN estimates were obtained with different equations.

warnings (see Table 4.1). Because *WARNAREA* is a linear combination of the labeling format dummy variables, it is not feasible to include both in the same equation.

The general character of the results in Table 4.7 parallels that for the individual label variables. The probability that the consumer will take precautions increases in all cases as the value of *WARNAREA* rises. This result is consistent with the theory in Section 4.1 (see equations 4.7 and 4.11). Given that consumers of the two products generally underestimate the risks (that is, $s > 1$), more risk information induces more precautionary effort. Only for the precaution to avoid mixing bleach with ammonia-based cleaners is the effect not statistically significant (at the 5 percent level).

Although it is likely that all of the hazard warning information provided will influence consumers' risk perceptions at least to some extent, we expected the information specifically targeted to a particular precaution to be most influential. To analyze this possibility, we developed a variable *SPECWARN*, which is the fraction of the labeling area devoted to each of the specific precautions. The *SPECWARN* variable is statistically significant in four of five cases, and in all instances either *WARNAREA* or *SPECWARN* has a statistically significant effect on precautions.

The magnitude of the effects also accords with the theoretical predictions. Providing additional specific precautionary information should have greater informational content with respect to that precaution than an equivalent amount of label area for general hazard warnings. The magnitudes of the *SPECWARN* coefficients consequently should be greater than the magnitudes of the *WARNAREA* coefficients. This relative impact is reflected in four of the five coefficients in Table 4.7.[5] The relative discrepancy in the impacts is greatest for the bleach precautions relating to mixing the product with toilet bowl cleaner or ammonia-based cleaners. Unlike other precautions, such as the need to store the product out of the reach of children, the precautions relating to the risk of chloramine gas poisoning are particularly difficult to infer from any general hazard warnings. The substantial differential impact consequently accords with our expectations.

It is also instructive to compare the precaution that was common to both labels (storing in a childproof location). The *WARNAREA* and *SPECWARN* coefficients for childproofing in the bleach equations are almost double those for drain opener. This result is consistent with the earlier findings. Although the child poisoning risk from drain opener is more severe and should create a larger impact, this influence is mitigated by differences in format. The child poisoning warning is not as prominent on the drain opener label, which treats child poisonings as one of many hazards. In contrast, the bleach label emphasizes this particular warning. Overall, the results presented here and earlier are consistent with the hypothesis that precautionary behavior will be influenced by the provision of risk information and the amount of risk information, which is governed both by the amount of hazard information given and by the format used to convey it.

4.4 The Value of Precautionary Actions

Although the information presented on the warning labels had many of the expected effects, there is no assurance that consumers' subsequent behavior was completely consistent with the model of Section 4.1. Many consumers may not undertake precautions even in the presence of the warning. Failure to take precautionary actions is not necessarily irrational, since precautions are costly, but it is instructive to ascertain whether this behavior seems to be irrational in view of the associated cost of precautionary behavior. The other danger is that labels may be unduly alarmist, leading consumers to undertake precautions that are not worthwhile. That possibility will also be explored.

Protective Actions and the Disutility of Taking Precautions

Rational consumers undertake safety precautions only if the expected value of the safety gains exceeds the associated disutility of the precautionary actions. Because our survey included questions pertaining to this disutility, in conjunction with assumptions about the associated risk we can calculate the critical value of the health outcome that is required to lead consumers to undertake each particular precaution.

Faced with a binary choice of whether or not to take precautions, consumers choose to exercise care if the associated disutility V is below the value of the risk reduction, which is the product of the change in the risk p associated with precautions and the value of the health loss L. For all consumers for whom L exceeds $-V/\Delta p$, precautions are desirable. Since detailed information on the heterogeneity of risk levels is unavailable, we cannot make a calculation of this type on an individual basis, but we can derive suggestive results by assuming that individuals can be characterized by the average risk in the population.

The first component in the analysis is the disutility associated with the precautions. Our survey included questions about consumers' willingness to pay per bottle for a variety of product characteristics ranging from a fresh lemony scent to attributes related to the disutility of taking precautions, such as the need to wear gloves when using the product. To avoid biasing subjects' responses about the disutility of taking precautions, we phrased the questions so that respondents would not take into consideration the risk reductions achieved by those precautions.

Column 1 in Table 4.8 reports the sample's mean willingness to pay per bottle to avoid the need for taking each of the precautionary actions, expressed in increases in current price ($.79 for bleach and $1.79 for drain opener). The average product price increases ranged from $.15 to $.19 per bottle. This fairly narrow range in valuations may falsely suggest that consumers did not attempt to distinguish their underlying preferences but instead gave uniform responses of $.10 or $.20 to all questions. As is documented in Appendix C, the valuations any consumer expressed for different product characteristics were not strongly correlated. Rather, there is much more variation when the other product attributes included in the survey

Table 4.8 Precaution-related decision components

Precaution	1 Mean disutility in dollars of precaution per bottle (std. dev.)	2 Mean number of containers used annually	3 Nature of risk	4 Annual household risk without precautions	5 Critical benefits value ($)
Do not mix bleach with ammonia-based products or toilet bowl cleaner	0.19 (0.46)	12.2	chloramine gas poisoning	0.000058	37,900
Store bleach to prevent access by children	0.16 (0.46)	12.2	nausea and stomach cramps for one day	0.000061	32,000
Wear gloves to prevent drain opener hand burns	0.17 (0.34)	1.78	temporary hand burns	0.000061	5,200
Store drain opener to prevent access by children	0.15 (0.33)	1.78	very severe internal burns, possibly irreversible	0.000041	6,500

(such as using the cap as a measure) are considered. As a result, we were able to reject the hypothesis that consumers gave uniform responses to all product attribute questions. To complete the calculation of the precaution's associated annual disutility, we multiplied the disutility per bottle by the number of bottles used per year (shown in column 2 of the table).

Columns 3 and 4 list the kind of injury associated with each precaution and the average household risk if the consumer does not take precautions. This risk figure was calculated from information on total poisonings (from the U.S. Poison Control Centers and Consumer Product Safety Commission), coupled with information about the fraction of consumers who took precautions with current labels. To facilitate the calculations, we assumed that taking precautions would reduce the risk to zero. The incremental risk reduction achieved in each case is rather small: all the annual household risks are below 0.0001.

The final column in Table 4.8 reports the critical valuation of the health outcome that would be needed for consumers to find it economically desirable to take precautions. Under our assumption that the average figures for risk and disutility characterize all consumers, individuals who take precautions have valuations above the critical amount and those who do not take precautions have health loss valuations below the critical amount. For the bleach risks, precautions are desirable if the value of avoiding a chloramine gas poisoning is at least $37,900 and the child poisoning valuation is more than $32,000. These values are sufficiently high in view of the generally temporary nature of the ailments that it is easy to envision consumers who would rationally choose not to take these precautions.

The critical valuations for the drain opener health outcomes are lower, largely because fewer bottles of this product are used annually. If consumers value avoiding hand burns at $5,200 or more and child poisonings at $6,500 or more, precautions are desirable. Consumers' hand burn valuations may exceed or fall below these critical cutoff levels, so some mix of responses is to be expected in this case.

The most striking result is that for the drain opener poisoning risk to children, which is by far the most severe outcome, the critical value of the loss that is needed before precautions are desirable is almost the lowest in the table. Therefore, we would have expected consumers to undertake the associated precaution most often in this case. No such dramatic difference was observed, and in many cases other precautionary warnings were more effective. Indeed, after reading labels with the hazard warning, more households with children under age five would store bleach in a childproof location than they would drain opener (Table 4.6).

This result reinforces the earlier findings regarding the childproofing warnings on the drain opener labels. By failing to communicate the need for precautions in a prominent fashion the drain opener labels led consumers to take actions that clearly are not plausible in view of the desirability of the precaution and the merits of exercising care relative to other products. Although the overall nature of the behavioral responses is consistent with rational behavior, fully consistent results clearly were not achieved. This difficulty seems largely attributable to the character of informational transfer, but shortcomings in individual decisionmaking capabilities cannot be ruled out as a contributing factor.

Other Product Uses

If labels are unduly alarmist, they might affect consumers' incentive to undertake other forms of precautionary behavior. To test for this possibility, we included in the questionnaire other questions about the proper use of the product. For bleach these questions were not risk related, but for drain opener they were. The principal results, summarized below, are reported in greater detail in Appendix D.

Two fundamental uses of bleach are to remove mildew and to use in a wash for problem stains. These uses were printed on all four bleach product labels, and there was no difference in the effect of the labels on these actions.

Other product uses described in a more selective fashion also performed as expected. The label with no warning and the Clorox label specifically indicated that the cleaning product could be used to clean sinks and floors, whereas the Bright and Test labels did not include this message. Consumers were significantly less likely to undertake such uses of the product in the case of the Bright and Test labels. This result suggests that the usage information was processed reliably and was not distorted by the hazard warnings.

We found much more interdependence between the risk information and the usage information on the drain opener label for which several usage directions were risk related. All of the labels, including the one with no warning, advised against pouring the product through standing water, and we observed no difference in their effectiveness. We obtained similar results with respect to whether or not the consumer believed that the product could be used with a septic tank. This also was an expected result, since there is no reason to avoid such use. Finally, the Drāno and Test labels advised against the use of a plunger with the drain opener, but the additional precautions that would have been undertaken were not statistically significant. This precaution was of subsidiary importance in terms of both the label and consumer practices: because few consumers envisioned the use of a plunger, the potential labeling effect was limited.

Overall, the evidence suggests that the diverse information contained in the label was processed reliably. Moreover, there is no clear-cut evidence that the labels distorted other usage decisions.

4.5 Implications

Although information provision policies have considerable appeal on conceptual grounds, their efficacy has long been questioned. Informational policies have typically been undertaken in situations in which individual decisions are not believed to be fully rational because of failure to understand fully the risks associated with different actions. Although informational policies can potentially address an inadequacy of individual knowledge, if the inadequacy stems from an inability to process risk information reliably, these policies will not be effective. Moreover, if decisions involving low-probability events are fundamentally flawed by limitations on individual rationality, informational approaches also may not remedy these difficulties.

The experimental results presented here for a sample of consumers provide a more optimistic view of the potential efficacy of informational approaches in affecting precautionary behavior. Consumers responded in a manner that was broadly consistent with the main predictions of an economic model of rational safety-related actions. Households facing particularly large risks were more likely to undertake protective actions, and differences in the information provided produced the expected effects.

The overall efficacy of informational approaches will be governed not only by the level of the risk conveyed, but also by informational content. Many educational campaigns have exhibited disappointing results because their informational content has been low. Some widely used consumer product labels examined in our survey likewise had negligible effects in some instances because they did not provide risk information in an effective manner.

Labels will not lead all consumers to take precautions, because for some of them the disutility of taking the precautions outweighs the value of the benefits of reduced risk. Analysis of the implicit values of the health outcome associated with precautionary actions is helpful in highlighting potential shortcomings in informational strategies or individual behavior. In the case of drain opener storage to prevent access by children, it was clear that precaution-taking either fell short of the optimal amount or else that consumers overreacted to the less severe risks on other labels. This deficiency could be ameliorated through improvement of the label's structure to provide the risk information more prominently.

Informational strategies will not necessarily lead to optimal results after such changes. The most that can be concluded at this point is that information can produce precautionary behavior consistent with the most salient predictions of rational economic actions.

5 | Risk-Dollar Tradeoffs, Risk Perceptions, and Consumer Behavior

WESLEY A. MAGAT,
W. KIP VISCUSI, and
JOEL HUBER

Psychologists, economists, and decision theorists have long been interested in how consumers make decisions under uncertainty (for example, Kahneman, Slovic, and Tversky 1982; Arrow 1982). If these decisions are assumed to be the result of optimizing behavior, a fundamental concern is the rate of tradeoff between different valued attributes. Using the same household chemical products described in Chapters 3 and 4, we focus on consumers' tradeoffs between two attributes of general concern—the risks of using the products and their price. Unlike the results in Chapter 4 on the risk–precautionary effort tradeoff, the results on risk–dollar trade-offs can be more readily compared with other estimates in the literature to assess consumer rationality. Indeed, it was by analyzing the initial risk-dollar tradeoffs needed for undertaking precautions that we were best able to analyze the consistency of individual choices in Chapter 4.

The advantage of putting the tradeoff in terms of a risk-dollar metric rather than a risk–safety effort metric is that the dollar scale is well ordered and comparable across individuals. In contrast, we do not know whether wearing gloves is more onerous than keeping bleach in a childproof location; nor do we know how such valuations differ across consumers. In this chapter we present results on the risk–dollar tradeoff obtained directly through the contingent valuation and conjoint questions. These results are then used to evaluate consumers' implicit valuations of the health risks posed by the product. Although we can provide some general empirical reference points for the plausibility of the results, there is no existing study of these particular health risk valuations with which our results can be directly compared. The paucity of meaningful empirical reference points stems from the comparatively undeveloped nature of the morbidity valuation literature, which is much less extensive than the literature on the value of life.[1]

Our consumer study examines two acute morbidity risks from the use of each hazardous chemical product. In addition, we consider individual risks from using the products, in contrast to the problem of valuing public goods (such as air pollution reduction) or common property resources (such as use of a goose population for recreational hunting). Until now most of the literature on nonmarket valuation of environmental amenities has focused on nonprivate goods (see Cummings, Brookshire, and Schulze 1984; Bishop and Heberlein 1979).

Finally, our study requires consumers to value low-probability risks. The growing literature on this subject suggests that consumers have particular difficulty thinking about low-probability events. Our results suggest both the existence of a bias introduced by the low levels of the probabilities of risk and an explanation for that bias.

We address four specific valuation issues concerning the consumer choice problem described above. First, how does one measure risk-dollar tradeoffs? We employ two different approaches to this problem, the direct contingent valuation method and the use of paired comparison, or conjoint questions. Our application of these approaches also illustrates an innovative method of administering the questionnaires with personal computers.

Second, how do the estimated morbidity values differ from morbidity and mortality values estimated from market-based behavior, such as labor market activity? If our consumer survey estimates are substantially different from market-based morbidity values, the consistency and overall rationality of the choices will be in question.

Third, how do the low values of the probabilities of injury from the products affect the values the consumers attach to avoiding those risks? Finally, what is the relationship between the actual risks of injury from the chemical products and consumers' perceptions of those risks?

Chapter 4 explained the link between the content and the format of the information on product labels and the extent to which those labels induce consumers to take precautions. Both an individual consumer deciding whether to take the precautions and a policymaker contemplating issuing labeling requirements must be able to compare, in some common units, the costs of the disutility of precaution-taking with its benefits. Assigning the morbidity benefits and disutility costs a dollar value allows them to be compared. (For further discussion of this point and an illustrative cost-benefit analysis, see Viscusi and Magat 1985.)

The next section describes the details of the survey design used to elicit the risk-dollar tradeoffs underlying individual decisions. It explains both the contingent valuation and paired comparison methodologies. After presenting the results of the morbidity estimates in Section 5.2, we compare those estimates with morbidity and mortality estimates derived from market responses. Section 5.3 also offers an explanation for the biases observed in the low-probability risks. The final two sections present data on consumer risk perceptions and discuss the meaning of our results.

5.1 Survey Design

Our results are based on the responses to the stage 6 questions in both the cleaning agent and the drain opener questionnaires. Since the subjects had completed all the questions about precaution-taking that would be sensitive to whether the cleaning agent was identified as a bleach, we referred to it as bleach in this stage.

The interviewer began stage 6 by explaining to the subjects that they would be asked to respond to two more sets of questions from a personal computer. For subjects receiving the household bleach, the computer program first described the health consequences of chloramine gas poisoning created by mixing bleach with another product containing ammonia and then asked questions designed to reveal consumers' willingness to pay to avoid chloramine gas poisonings. After describing the dangers from ingestion of bleach by a child, the second set of questions sought to determine the value subjects attached to reducing the incidence of child poisonings. Subjects receiving the drain opener were asked two similar sets of questions about two of its major hazards, hand burns and child poisonings.

```
        CURRENT BLEACH                          NEW BLEACH
*****************************        *****************************
*                           *        *                           *
*  Cost per year:           *        *  Cost per year:           *
*        $10.00             *        *        ?                  *
*                           *        *                           *
*  Injury level:            *        *  Injury level:            *
*                           *        *     50% DECREASE in        *
*     50  gas               *        *     gas poisonings         *
*     poisonings for        *        *     compared to the        *
*     every 2,000,000 homes *        *     current product        *
*****************************        *****************************
```

How high would the price of the NEW BLEACH have to be before you
would rather buy the CURRENT BLEACH?

$_____/year

PRESS THE NUMBERS THAT SHOW HOW MANY DOLLARS AND CENTS YOU ARE
WILLING TO PAY.

Figure 5.1 Sample contingent valuation question.

As described in Chapter 3, we used two methods to elicit the subjects' morbidity valuations—the contingent valuation (CV) approach and an approach based on paired comparisons of products. The contingent valuation approach has been extensively discussed over the past decade, as analysts have attempted to utilize survey data to value environmental amenities (for example, Desvousges, Smith, and McGivney 1983; Cummings, Brookshire and Schulze 1984; Desvousges, Smith, and Freeman 1985). Survey and experimental methods are widely used in a host of behavioral areas in psychology.

The CV method typically obtains either a series of bids ending with the maximum willingness to pay for the environmental benefit or asks one direct question about maximum willingness to pay. We chose the latter approach. Figure 5.1 shows our CV question for valuing reductions in the chloramine gas poisoning rate from using bleach. It presents two products (for example, a current bleach and a new bleach) that differ by only two characteristics—cost per year and the number of injuries for every 2 million homes (which subjects were told is the number of households in North Carolina). For a given percentage reduction in the injury level associated with the new product, the CV question asks how high the cost of the new product must be before the subject would rather buy the current (riskier and less expensive) product. We then calculated the implied value per injury avoided (in every 2 million households) by dividing the difference in annual costs for the two products by the number of injuries avoided.

In contrast to the one-step CV procedure, the paired comparison approach asks each subject to make a series of comparisons of pairs of products with differing characteristics. This approach is widely used in marketing research because it poses the issue in terms of the context that reflects actual product choices in the marketplace. As shown in Figure 5.2, for each pair of products, the current product remains unchanged, but the cost and injury levels for the new product change. Subjects were asked to rate on a scale of one to nine which of the two products they preferred, with five representing indifference between them. From the five rankings of paired comparisons presented to each subject, we used two techniques to estimate subjects' willingness to pay to avoid accidents: conjoint analysis of each individual's responses and regression analysis of the pooled responses for which the subjects were indifferent. (The next section describes these two techniques in more detail.)

The CV and paired comparison responses were elicited through questions asked on an IBM Personal Computer. The interviewer sat next to the subject to answer questions if the subject did not fully understand any questions posed by the computer, but such questions rarely arose.

Besides the obvious advantages from accurately storing and retrieving subject response data on the personal computer, this computer-based interview approach reduces the potential for interviewer

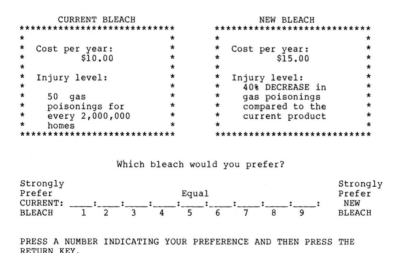

Figure 5.2 Sample paired comparison question.

bias because the program asks every subject the questions in exactly the same way. In addition, there is less incentive for respondents to misrepresent their responses so as to make a more favorable impression on an interviewer.

Perhaps the most innovative aspect of our paired comparison technique is its interactive nature. The computer program utilized the subject's responses to the first two paired comparison questions to design the tradeoffs posed in the later questions. The program estimated a linear version of the subject's rate of tradeoff between money and injuries from the first two questions and used this estimate to design the subsequent three paired comparisons. The criterion for this design was that the pairs be in the neighborhood of the subject's likely range of indifference between the two products. This ensured that for the subsequent comparisons the subject would not strongly prefer either one or the other. This adaptive design reduces the number of paired comparisons that each subject must answer in order to produce stable estimates of the tradeoff between money and injury reduction. The design also greatly diminishes the possibility of bias from ceiling effects caused by subjects forced to use a "1" or a "9" ranking for strongly one-sided choices.

5.2 Injury Valuations

Table 5.1 displays estimates of the valuations of the four injuries from both the CV responses and the paired comparison responses. Although about half the subjects answered a CV question first and then the paired comparison questions, and the other half answered the paired comparison questions first, for comparison purposes only the first set of responses made by a subject is shown. The answers to the CV questions indicate how much more the respondents were willing to pay for a new, safer product. The response represents the value of reducing the probability of the injury by the designated percentage (such as 50 percent for the example in Figure 5.1). Given the annual number of injuries associated with the current product per 2 million homes, we calculated the number of injuries avoided and divided the indicated cost increase by that number to derive a measure of the subject's implied value per injury avoided (in every 2 million households).

Because the distributions contained some extreme outliers, we excluded the top five observations and the bottom five observations

Table 5.1 Comparison of injury valuations[a]

Injury type	(1) Mean CV[b]	Paired comparison data		
		(2) Mean conjoint	(3) Mean from indifferent pairs data	(4) Regression of indifferent pairs: restricted sample
Bleach				
Chloramine gas	0.15	0.69	0.72	0.77
	(0.02)	(0.11)	(0.09)	(0.09)
Child poisoning	0.21	0.25	0.35	0.29
	(0.03)	(0.03)	(0.04)	(0.03)
Drain opener				
Hand burns	0.06	0.62	0.55	0.55
	(0.01)	(0.08)	(0.06)	(0.06)
Child poisoning	0.18	0.41	0.55	0.50
	(0.02)	(0.10)	(0.06)	(0.05)

a. Injury valuations are measured in dollars per injury avoided in every 2 million households. Standard errors are shown in parentheses.

b. The highest five and lowest five injury valuations were excluded before calculation of the mean values.

in each distribution. Although this procedure makes the means somewhat arbitrary, we believe that the truncated means provide a more useful summary statistic than nontruncated means for the distribution of values for each of the four groups. Column 1 of Table 5.1 lists the truncated mean values of these morbidity values derived from the CV responses for subjects who answered the CV question before the paired comparison questions.

Conjoint analysis has been used extensively in the marketing literature (Green and Srinivasan 1978) to estimate the structure of consumers' relative preferences for different product attributes from ratings of paired comparisons. (See Appendix E for a discussion and test of the linearity of the rating scale.) We used this approach to recover each subject's relative preferences for money and risks, regressing the rating observations ($RATING$) for each of the five product pairs against both the change in cost levels ($\Delta COST$) for each pair and the change in injury levels ($\Delta RISK$) for each pair. The results provide a set of regression estimates (coefficients α and β) of the relative importance of money to injuries for that consumer, or

(5.1) $$RATING = \alpha \cdot \Delta COST + \beta \cdot \Delta RISK.$$

By taking the absolute value of the ratio of the two coefficients ($|\beta/\alpha|$), we calculate each subject's willingness to pay to reduce the accident rate by one accident per 2 million households. Column 2 of Table 5.1 displays the mean responses for the morbidity valuations derived from the conjoint approach.

As a check on the accuracy of the conjoint approach, we performed alternative estimates of the injury valuations that avoid the need for a rating scale, restricting our attention to only those pairs of products for which a subject was indifferent. Column 3 gives the mean value (for all subjects' indifferent pairs) of the ratio of the cost difference to the risk difference for each pair of products. Column 4 presents the coefficient of the variable, the difference in risk ($\Delta RISK$), in a regression of the difference in cost ($\Delta COST$) against the difference in risk.[2] The injury valuations in columns 3 and 4 fit closely with those derived from the conjoint approach in column 2. This congruence supports the validity of the conjoint methodology for measuring the injury valuations implied by responses to paired comparison questions.

The statistics in Table 5.1 show fairly convincingly that the direct contingent valuation approach to eliciting injury valuations yields substantailly *lower* values than the use of paired-comparison responses. The subsample of indifferent pairs produces estimates close to those derived by using conjoint analysis on the full sample of paired comparison responses, and all the estimates from the paired comparison approach exceed the CV estimate for each of the four injuries.[3]

5.3 Low-Probability Bias

Although so far our analysis has focused on the differences in the responses to the two survey techniques, our ultimate objective in eliciting these figures was to assess consumers' willingness to pay to avoid adverse health outcomes. These findings are of potential interest for purposes of benefit assessment, but more fundamentally they pertain to how individuals make choices under uncertainty. One of the major contributions of this study may be to suggest what kind of decision heuristics consumers use in risky choice situations.

Several researchers have argued that behavior involving low-probability events differ from their decisions involving higher-probability events by amounts that are not consistent with the differences in probabilities. Theoretical and empirical arguments have been generated to support a bias in both directions—toward overweighting low-probability events as well as toward underweighting them.

Kahneman and Tversky (1979) have posited that consumers may use a probability weighting function that overweights low probabilities to weigh the values realized under different states of the world. This is the pattern observed, for example, in the study of consumer mortality risk assessment discussed in Chapter 1. In addition, casual observation suggests that local residents and consumers overresponded to the risks imposed by the 1979 Three Mile Island nuclear power plant accident and the 1981 Tylenol poisonings. In a contingent valuation study of consumer willingness to pay to avoid the risks from hazardous wastes that was concurrent with our own, Smith, Desvousges, and Freeman (1985) have found that individuals do in fact attach a higher marginal valuation to risk reduction in cases of low-probability accidents than to higher-probability ones.

In contrast to this overreaction hypothesis, Kunreuther et al. (1978) have argued that even for catastrophic events, when the probabilities are so low as to fall below a threshold, consumers essentially ignore the events by treating the probabilities as if they were zero. Their extensive evidence is based on reactions to flood and earthquake hazards.

One hypothesis that is consistent with both the overreaction and underreaction to low-probability events may be the inability of individuals mentally to trade off probabilistic characteristics of different events. By focusing on either the magnitude of the loss or the probability of its occurrence and ignoring the second characteristic, they tend either to overreact to the risky event or to ignore it. If people have difficulty internalizing low probabilities but are *forced* by a survey to consider them in making decisions, they may respond by mentally augmenting the probability to a level that is familiar to them. In contrast, if decisionmakers are allowed to ignore low-probability events, as in making actual flood insurance decisions, they may do so in order to simplify the processing costs of making those decisions.

The mean CV values in column 1 of Table 5.1 describe the willingness to pay for a one-injury reduction for every 2 million households rather than an individual morbidity value. To derive the value of avoiding each "statistical injury" it is necessary to multiply the figures in column 1 by 2 million. The resulting values per injury avoided are $300,000 for gassings from bleach, $420,000 for child poisonings from bleach, $120,000 for hand burns from drain opener, and $360,000 for child poisonings from drain opener.

The absolute levels of these morbidity valuations appear to be excessive. It seems implausible, for example, to assign an implicit value of $120,000—or four times the respondents' average household income—to a hand burn that will have a minor impact on health. These results are also out of line with the results of hedonic wage studies (surveyed by Viscusi 1986). The estimated values of a job injury range from $20,000 to $30,000—roughly an order of magnitude below the value placed on the morbidity effects in the survey. The bleach and drain cleaner risk valuations are more in line with estimated implicit values of life, which are on the order of $650,000 for workers in high-risk jobs and $3.5 million for workers in average-risk jobs. In short, the morbidity survey results cannot be reconciled with related empirical assessments of health effects.[4]

A possible explanation for this phenomenon is that the subjects may not have fully understood the probabilities involved. For example, the subjects may have focused their attention much more on the numerical or percentage reduction in injuries between the two products in any pair, without closely internalizing the base number of households (2 million) subject to risk. Had a base of 200,000 households resulted in about the same responses to the CV and paired comparison questions, then the implied total values of injury reductions would have been ten times less if the respondents' dollar valuations of the risk reduction had not changed. If there is lack of attention paid to the base number of households, this phenomenon could be explained by either the decision process simplification strategy of using primarily one piece of data (in this case the numerical or percentage reduction in injuries) or an inability to comprehend rates generated from such large denominators. There is no reason to believe that telling respondents that their risk is 0.00001 annually would have been any more effective.

These results suggest that individuals have great difficulty in making decisions about situations involving low-probability events. In particular, they may be quite insensitive to probability base numbers. Their "errors" in behavior appear to stem partially from a large random component, but also from a consistent overreaction to quite small risks. These results are in keeping with the other findings in the literature indicating that when decisionmakers are forced to make choices involving low-probability events, they overstate their true probabilities and thus overvalue them relative to events with higher probabilities.

Another reason for the difference between our findings and those of Kunreuther et al. (1978) is that our analysis focused on health risks, whereas theirs focused on financial risks from disasters. As they have noted in their study, consumer responses may vary with the kinds of risks involved even though their welfare implications are the same. In their recent review of the literature on the psychology of risk perception, Fischhoff, Slovic, and Lichtenstein (1980) have noted that risk perceptions are influenced by many aspects of the risk, such as the immediacy of the hazard and whether the outcome leads to visible and sensational outcomes, as in the case of tornadoes.

To the extent that health risks generate stronger reactions than equivalent monetary risks, our results tend to yield higher risk-dollar tradeoffs than in the disaster insurance study by Kunreuther et al. From an economic standpoint, however, it is not possible to ascertain whether our consumers' high valuation of health risks stems from an overreaction to risks of a health-related nature that would not have been evident for equivalent monetary risks or whether in fact consumers do associate a higher welfare loss with these health effects.

The most that can be concluded at this stage is that there is clear evidence of consumer overreaction to small health risks. Whether such a result also generalizes to financial risks is a matter for future study. It is, however, noteworthy that for the job risks considered in the following chapter, which are health related but larger in magnitude than the consumer risks, there was no evidence of overreaction. On the basis of this evidence, the level of probability rather than the nature of the risk appears to be the primary contributor to our results.

5.4 Consumer Risk Perceptions

A key variable influencing whether or not individuals undertake precautionary actions is the risk they believe to be associated with the product. This risk assessment is also pertinent in establishing the household-specific risk to the extent that individuals have better risk judgments than could be obtained from objective injury data.

To ascertain the value of individuals' risk assessments for the four injuries in this study, we asked respondents to scale particular risks relative to those faced by an average household. Another procedure that we considered was to present respondents with a linear risk scale, which they would then mark to indicate their true risk, as in our worker study in Chapter 6. But because of the low level of risk involved—on the order of one accident per 50,000 households, or one accident per 150,000 people—and the difficulty of finding meaningful reference points to enable consumers to make such refined distinctions, we decided instead to inform consumers of the true average risk and then to ask them whether or not they believed they faced above-average risk, below-average risk, or the same risk as the average. For example, in the case of hand burns from drain openers, we gave respondents the following information: "If liquid drain opener splashes on to someone's hands, it causes painful burns or red, swollen blisters. The treatment is to see a doctor, who will carefully wash the injured hands. Complete healing occurs within a week. This accident occurs to 45 out of 2,000,000 households (the number of households in North Carolina) which use drain opener. Of course, this accident is more or less likely to happen in different homes depending on how liquid drain opener is used. Suppose you used this product in YOUR home. Would the chance of an accidental hand burn be more or less likely in your home than in an average home that uses liquid drain opener?"

Subjects who indicated that their household's risk was above average were then asked to rate how much more likely the risk was by either picking one of five different options (as likely, twice as likely, 4 times as likely, 10 times as likely, 100 times as likely) or picking an alternative number that better expressed their relative risk. The wording was similar for accidents rated as less likely than average.

We followed the same procedure in the case of child poisonings for drain openers, for which the risk per household was 35 injuries out of every 2 million households. Consumers were also apprised of the greater incidence of this risk among children under five years of age. The risk for chloramine gas poisoning from bleach was given to be 50 out of every 2 million households experiencing the accident, and the risk of child poisoning from bleach was given as 70 out of every 2 million households. Once again, in the case of child poisoning, respondents were informed that the risk was principally to children under the age of five and were told the nature of the risk (such as temporary stomachaches). The risk figures were based on available statistics from the National Clearinghouse for Poison

Table 5.2 Distribution of relative risk assessments (percentages)

Risk	Above average	Average	Below average
Drain opener			
Hand burns	3	46	51
Child poisoning	3	32	65
Bleach			
Chloramine gas	3	50	46
Child poisoning	2	57	40

Control Centers and from the Consumer Product Safety Commission.

The results in Table 5.2 are quite striking. In all cases very few consumers—3 percent or less—consider their household to be above average in risk. Roughly half consider their households to be about average in risk, and the other half below average. It is noteworthy that the largest share—65 percent—view their households as being below average in risk for child poisonings from drain openers. This is by far the most severe risk in the sample, as burns to the throat may cause permanent loss of use of the esophagus. Consumers appear to be particularly optimistic that this very adverse advent will not occur.

The high degree of optimism among consumers is not unprecedented in studies of risk perception. Interviews on automobile driving behavior by Svenson (1979) have suggested that most drivers view themselves as being among the most skillful and safe drivers in the population. Similarly, a study of consumer risks by Rethans (1979) has found that 97 percent of consumers believe that they are either average or above average in their ability to avoid accidents from bicycles and power mowers. This high fraction of people who believe that they are relatively safe suggests that there may be some overoptimism in individuals' risk perception when these questions are asked in relative terms.

The framing of the risk assessment questions may be an important factor. In Chapter 6, when we asked workers to rate the risks of their jobs in the chemical industry on an objective linear scale, the average risk assessment was in excess of the U.S. Bureau of Labor Statistics (BLS) injury and illness rate for the industry. After we removed the long-term health hazards by advising the workers that the chemical risks of their jobs had been eliminated and re-placed by a nontoxic chemical, the average risk perception was identical with the BLS injury and illness rate for the chemical industry. Similarly, Lichtenstein et al. (1978) have found that individuals overassess small mortality risks and underassess large mortality risks—a result that is shown in Chapter 6 to be consistent with a rational Bayesian learning process. These studies did *not* find systematic overoptimism in risk perceptions. As a result, the risk perceptions that actually drive individual behavior may not necessarily be biased, but when asked to make relative risk judgments, consumers may have an overoptimistic perception of their *relative* ability to avoid the risk.

5.5 Conclusion

While the results of Chapter 4 showed that consumers respond to labels in close to the manner predicted by our Bayesian model of learning and information acquisition, this chapter's results suggest that the tradeoffs underlying these decisions may be biased in a systematic manner. In particular, consumers appear to overstate their dollar valuations of their willingness to pay to avoid morbidity risks for hazardous chemical products, at least relative to morbidity and mortality values implicit in market-based behavior such as in labor markets.

We hypothesize an explanation for this result based on individuals' difficulties in mentally trading off the magnitude of a loss and its probability of occurrence. In choice situations in which consumers are forced to focus on the probability of rare injury occurring, they may overestimate the probability, implying a higher willingness to pay to avoid the injury.

The implications of this finding for the sales of hazardous consumer products are of great consequence to their sellers in terms of effective label design. If labels or other information programs are constructed to force buyers to focus on the risks of injury from a product, no matter how small (as long as they are positive), then many consumers will avoid purchasing the product even though they regularly accept similar risks in other activities and in other consumption decisions. In contrast, if information policies, such as labeling, are designed to hide the risks of injury from a product, consumers may ignore the risks when making their purchasing decisions.

The second class of results pertaining to biases in *relative* risk perceptions may be a consequence of the framing of the survey question in relative terms. Because our subjects' underestimation of the risks is not consistent with other evidence of consumer overestimation of low-probability events, it is not at all clear whether consumers overestimate or underestimate the risks they face. All that we can be confident of at this stage of research is that thinking about low-probability events imposes considerable demands on individuals' cognitive capabilities, and it is unlikely that the behavior that results will be ideal.

6 | Hazard Warnings for Workplace Risks: Effects on Risk Perceptions, Wage Rates, and Turnover

W. KIP VISCUSI and CHARLES O'CONNOR

The provision of hazard warnings is by no means restricted to contexts involving consumer choice. Firms have long provided hazard warnings of various kinds, and the Reagan administration's requirement that hazardous chemicals be labeled is the most expensive social regulation that it has issued to date.[1] The issues involved are quite similar to those in consumer contexts. In each case, the fundamental issue is how individuals will process the information and how it will ultimately affect their risk-related decisions.

Examination of the effect of information on worker actions takes on added interest because of differences that do not stem from the consumer/worker choice distinction. The first dividend from examining workplace risks is that with a new survey for a different risk context one can examine other classes of issues. The results presented here are from a survey investigating the effect of hazard warnings on the level of workers' risk perceptions, reservation wage rates, and turnover behavior. Among the novel issues considered is whether the risk level per se of the warning or the warning's informational content is more consequential. In addition, because the survey addresses how workers are affected by the risks they now face as well as how they will respond to risks after seeing the hazard warning, we can analyze how information changes worker behavior. This structure of the survey is particularly advantageous from a statistical standpoint.

A second advantage of studying risk information in the workplace is the existence of a large literature analyzing the effects of job hazards on wages, turnover, and other economic concerns (see Viscusi 1979, 1983 and the references cited therein). This emphasis on worker risks rather than consumer hazards is largely a result of the greater availability of data on job risks. This extensive literature, with its variety of benchmarks for how workers respond to risks, provides a basis of comparison for the results involving the provision of job risk information.

On the other hand, research on information processing issues has been suggestive rather than conclusive. Viscusi (1979) has found that workers' risk perceptions are positively correlated with the industry risk and that workers who perceive job risks receive compensating wage differentials.[2] Nevertheless, workers in high-risk jobs display behavior consistent with an adaptive response: they accept jobs whose risks are not fully understood, learn about these risks from their on-the-job experiences, and then quit if these experiences are sufficiently unfavorable given the wage for the job. In effect, workers act as if they are engaged in an experimentation process.

Although the positive injury rate–quit rate linkage is consistent with an adaptive response, no study has investigated the dynamics of this relationship. Do workers learn about risks on the job, and does the resulting change in perceptions lead workers to revise their reservation wage rates (that is, the wage rates they require to work on the job) in the expected manner? More fundamentally, even in the absence of such learning, do workers have subjective risk assessments that generate compensating differentials in the manner that is consistent with studies of risk premiums for hazardous occupations and industries? In this chapter we extend this line of research by analyzing how workers make risk assessments, how they process information, and how changes in risk perceptions affect their decisions.

Since no existing data sets provide information on the evolution of workers' risk perceptions, we surveyed a sample of workers in four chemical plants to ascertain their responses to labels of potentially hazardous chemicals. We chose this form of information because the chemical industry already has some experience in designing chemical labels in an effective manner; thus it was possible to analyze the learning process rather than focusing on the design of the format for the information, as in the consumer study.

We first interviewed workers about the characteristics of their current job, such as their wage level and risk assessment. We then showed them a label for a hazardous chemical and told them that this chemical would replace the ones with which they now worked but that their jobs would otherwise remain unchanged. After providing this new information, we ascertained workers' new responses to the risk assessment questions, their likely incentives to quit the job, and similar concerns.

In Section 6.1 we describe the sample and discuss the results of the preinformation stage of the survey to establish a reference point for later results. These findings are also of interest in their own right because the survey provides extensive risk-related information, including detailed risk assessment questions and information on whether workers would repeat their job choice. These data enable us to establish a more direct link than in previous research between workers' risk perceptions and labor market outcomes, such as compensating differentials for risk. In Section 6.2 we discuss the chemical labels used and their effect on workers' risk perceptions. In Section 6.3 we estimate both the risk level implied by the hazard warning and the informational content relative to the workers' prior beliefs. This distinction between risk and informational content is of potentially fundamental practical consequence and has never been addressed empirically. The evidence presented is consistent with a Bayesian learning process in which workers retain some influence of their prior beliefs and incorporate the new information in the expected manner. In Section 6.4 we analyze the effect of risk information on compensating differentials and worker turnover. This analysis provides the first explicit test of the effect of changes in workers' risk perceptions on labor market performance.

The overall picture that emerges is that workers begin jobs with imperfect information but there are many rational elements to worker behavior, and the extent of risk-related mismatches of jobs and workers is not rampant. After acquiring risk information, most workers display the capacity to update their probabilistic beliefs in a manner broadly consistent with Bayesian analysis—that is, in a manner consistent with a rational learning process. Our observations of the adaptive responses to the new information suggest that workers are engaged in a continuous experimentation process in which they learn about the risks posed by their job and quit once the position becomes sufficiently unattractive.

6.1 The Sample and Baseline Results

Sample and Variables

Since no existing body of data provides longitudinal information on risk perceptions to make it possible to observe the effect of information on changes in risk assessments, we developed a questionnaire to enable us to analyze worker responses to job hazard information. (Appendix E reproduces the survey questionnaire.) Here we focus on the nature of the sample and the empirical results for the situation before workers received risk information. Because of the more comprehensive nature of the risk questions, it is possible to extend the empirical support for the principal labor market impacts of employment hazards to classes of behavior that have not been analyzed previously.

The sample consisted of 335 employees in the chemical industry. During the first six months of 1982 the managers responsible for chemical labeling interviewed workers at four plant locations of three major chemical firms. The operations at these plants included research and development as well as manufacturing. We chose workers in the chemical industry because potential changes in these individuals' jobs as a result of the labeling experiment would be of more than hypothetical interest. Chemical labels and changes in the chemicals with which they worked were routine elements of their current jobs.

The sample included a broad range of occupational groups exposed to chemicals: engineers, technicians, chemists, mechanics, researchers, and supervisors. Over half of the sample—185 workers—were either on hourly pay or were technicians. This group, which we call BC/TECH, closely parallels the blue-collar subsample analyzed in Viscusi (1979) and is the focus of many of the empirical results discussed in this section.

Table 6.1 summarizes the characteristics of the full sample and of the BC/TECH subsample. These characteristics followed the pattern to be expected for a national chemical firm. The average worker age was 39, and the majority of workers were white males (only 7 percent blacks and 43 percent females). The average education was two years of college, or 14 years of schooling. Almost two-thirds of the sample were married, and the average number of children (*KIDS*) was 1.36. Total worker experience (*EXPER*) was 18 years,

Table 6.1 Sample characteristics: Means and standard deviations[a]

Variable	Full sample ($n = 335$)	BC/TECH Subsample ($n = 185$)
AGE (years)	38.8	38.9
	(11.8)	(12.8)
BLACK	0.07	0.10
	(—)	(—)
MALE	0.57	0.42
	(—)	(—)
EDUC (years of schooling)	14.44	12.47
	(3.21)	(2.05)
MARRIED	0.64	0.62
	(—)	(—)
KIDS (number of children)	1.36	1.10
	(1.52)	(1.28)
EXPER (years of work experience)	18.38	19.19
	(11.68)	(13.0)
TENURE (years at firm)	8.19	7.15
	(7.22)	(6.41)
EARNG ($1982)	21,120.4	15,768.6
	(8,322.1)	(3,596.6)
DANGER	0.57	0.50
	(—)	(—)
RISK (scaled risk)	0.10	0.09
	(0.06)	(0.07)
HRISK	0.36	0.35
	(—)	(—)
WPREM	0.11	0.10
	(—)	(—)
TAKEA (repeat job choice)	0.79	0.77
	(—)	(—)
TAKEB (repeat job choice)	0.97	0.96
	(—)	(—)
QUITA (quit intention)	0.12	0.12
	(—)	(—)
QUITB (quit intention)	0.05	0.05
	(—)	(—)

a. Standard deviations for 0–1 dummy variables (d.v.) are omitted because they can be calculated from their fraction m in the sample, where the standard deviation is $(m - m^2)^{0.5}$.

8 of which were at the particular firm (*TENURE*). The average annual earnings (*EARNG*) were over $21,000.

The most distinctive characteristic of the survey is its series of risk perception questions. The *DANGER* variable pertains to whether or not workers felt their jobs exposed them to "physical danger or unhealthy conditions." Because the wording of this question parallels that in the University of Michigan (1975) Survey of Working Conditions used by Viscusi (1979), we can use it to assess the comparability of the empirical results. In the University of Michigan study, 52 percent of blue-collar workers viewed their job as dangerous. The results here are quite similar: 57 percent of the total sample viewed their jobs as dangerous, and 50 percent of the BC/TECH subsample perceived some risk. These strong similarities suggest that in the design of the survey instrument and the preinformation risk levels there is close comparability with the Survey of Working Conditions.

Although the mean *DANGER* levels are not unexpected, the relative riskiness rankings are, because the BC/TECH group presumably faced greater risks. Whether or not this assumption is correct is not clear: white-collar research chemists may in fact incur greater health risks than, say, maintenance personnel. The more similar results for the continuous *RISK* variable discussed below suggest, however, that these results may not stem from an actual difference in riskiness. Rather, the BC/TECH workers may have a less stringent risk level cutoff for considering whether their jobs are hazardous. Because willingness to accept a risk is negatively related to wealth, higher-income workers should be more likely to regard a job as dangerous, for any given risk level.[3]

Except in the case of one study using the *DANGER* variable, all previous analyses of risk premiums have used objective occupational or industry risk measures. For our study we developed a variable that would reflect workers' subjective assessments of the U.S. Bureau of Labor Statistics (BLS) injury and illness frequency rate for their jobs. This measure is a widely used index of industry hazards. From the standpoint of the theoretical foundations of the compensating differential theory, the wage-risk relationship should be driven by such subjective risk perceptions. Aggregative risk variables simply serve as an objective proxy for this variable when data on workers' actual risk assessments are unavailable.

To overcome the difficulties arising from different *DANGER* reference points and to provide a continuous risk measure that would permit a detailed analysis of worker learning, we developed a continuous *RISK* variable. We gave each worker a linear scale, ranging from very safe to dangerous. To provide an objective reference point, an arrow marked the average U.S. private-sector injury and illness rate. We then asked the subjects to mark on the scale the risk level they assessed for their jobs. This variable was then converted into probabilistic terms, that is, scaled between zero and one, where *RISK* is on a scale comparable to the BLS annual injury frequency rate. The mean *RISK* levels for the full sample and the BC/TECH subsample are comparable with the national average private sector risk probabilities and about 50 percent higher than the recent actual levels of the chemical industry's injury and illness frequency rate. This discrepancy is not unexpected. BLS statistics primarily capture safety-related accidents and underreport long-term illnesses from chemical exposures, and reported injury rates understate the actual risk level. Thus if their risk perceptions were accurate, chemical workers should "overassess" the risks relative to published risk statistics.

Using the *RISK* responses, we also created a job hazard dummy variable similar to *DANGER* except that the risk threshhold reference point was the same for all respondents. The high-risk variable, *HRISK*, assumed a value of one if the worker faced a risk above the U.S. average, and zero otherwise. A third of the sample viewed their jobs as being high risk, and two-thirds viewed their jobs as being comparatively safe. In conjunction with the earlier *RISK* results, these findings suggest that the chemical industry's relatively good accident record may be a reasonable reflection of most workers' perceptions, but the presence of substantial health risks leads a sizable minority to consider their jobs particularly hazardous.

Baseline Statistical Results

Since the time of Adam Smith, economists have observed that perceived risks will generate compensating wage differentials; that is, workers will demand extra compensation for jobs that pose extra risk (see Smith 1976, Thaler and Rosen 1976, Viscusi 1979, and Brown 1980). Table 6.2 summarizes the risk variable results for equations in which annual earnings (*EARNG*) and its natural logarithm (*LNEARNG*) serve as the dependent variables. Each equation also included an extensive group of variables that typically enter such earnings equations, such as the individual's education and work experience. For the BC/TECH subsample, the annual risk premium of $700–$800 for *DANGER* was of roughly the same mag-

Table 6.2 Compensating differential results[a]

Dependent variable	Sample	Risk variable	Risk coefficient (std. error)	Average annual risk premium ($1982)
EARNG	BC/TECH	DANGER	1,577.2 (438.1)	788.6
LNEARNG	BC/TECH	DANGER	0.097 (0.029)	746.5
EARNG	BC/TECH	RISK	6,898.4 (3,461.1)	636.2
LNEARNG	BC/TECH	RISK	0.479 (0.231)	665.3
EARNG	BC/TECH	HRISK	738.4 (465.5)	258.4
LNEARNG	BC/TECH	HRISK	0.053 (0.031)	289.8
EARNG	full (males)	DANGER	2,117.5 (775.6)	1,385.7
LNEARNG	full (males)	DANGER	0.124 (0.036)	1,875.3
EARNG	full (males)	WPREM	1,583.1 (1,179.9)	—[b]
LNEARNG	full (males)	WPREM	0.1094 (0.0549)	278.8
EARNG	full	DANGER	169.03 (529.51)	—[b]
LNEARNG	full	DANGER	0.018 (0.025)	—[b]

a. Each equation also includes the variables *AGE, BLACK, MALE, EDUC, MARRIED, KIDS*, and *EXPER*. The full sample results also include a *BC/TECH* dummy variable.

b. Annual risk premiums are not reported because the coefficients are not statistically significant (at the 5 percent level, one-tailed test).

nitude as the $900 annual compensation (1982 dollars) found for the blue-collar subsample in Viscusi (1979) for both *DANGER* and the BLS injury rate.

As with that study, the full sample results were not successful because of the impossibility of disentangling the wage premiums for risk from the positive overall relationship between job quality and individual income. The change in earnings equations in Section 6.3 are not subject to this difficulty. Restricting the sample to males only eliminates some of the problems arising from failing to control adequately for the omitted variables that determine individual earnings. Male workers' jobs tend to involve more direct handling of chemicals, and the annual risk premiums are considerably larger than for the BC/TECH subsample.

Of the three risk variables, *DANGER* yielded the largest annual risk premiums. These were somewhat larger than those for *RISK*, which were about $100 less. The above-average risk variable *HRISK* led to the smallest annual risk premiums, but the effects were consistently positive and statistically significant (at the 5 percent level, one-tailed test). This pattern may reflect the shortcomings of the *HRISK* variable, which may be a less accurate measure of the underlying job risk, thus leading to a downward bias in its coefficient. The general implications of these findings are less ambiguous. The consistently significant results from the subjective risk variables and the similarity of the *DANGER* and *RISK* premiums to those in earlier studies should bolster confidence in the validity of the compensating differential theory.

A closely related issue is whether workers are aware of any risk premiums. Since no previous study had asked workers whether they believed they received a risk premium, we developed a variable, *WPREM*, that assumed a value of one if workers believed they received higher pay because of the nature of the chemical industry and zero otherwise. This variable reflects compensating differentials for working in the chemical industry as opposed to some other industry, not risk premiums per se. Since two-thirds of the sample regarded their jobs as safer than the U.S. average, these incremental premiums should not be large. Only 10 percent of the sample believed they received such a chemical industry premium, and those that did earned an average wage premium of under $300, controlling for other factors (see *LNEARNG* equation, Table 6.2). As expected, the probability that the worker perceives a risk premium is strongly and positively related to each of the three risk variables, as the logit results in Table 6.3 indicate.

Since over one-third of the sample believed they faced above-average risks and only one-tenth acknowledged the existence of relative wage premiums, roughly one-quarter of the sample might appear to behave in a manner that is inconsistent with the standard theory. This need not be the case, however, since workers may, for example, earn some form of economic rent that makes the job attractive despite the absence of a perceived relative risk premium. Moreover, since the overall risk premiums average under $1,000 annually and only $300 for the relative chemical industry differential, many respondents may not have believed that the risk pre-

Table 6.3 Maximum likelihood estimates for perceived risk premium and turnover equations[a]

Dependent variable	Risk variable	Coefficient (asymptotic std. error)
WPREM	DANGER	2.96 (0.75)
WPREM	RISK	6.89 (2.78)
WPREM	HRISK	0.54 (0.38)
TAKEA	DANGER	−1.42 (0.35)
TAKEA	RISK	−11.22 (2.32)
TAKEA	HRISK	−1.53 (0.30)
QUITA	DANGER	1.21 (0.48)
QUITA	RISK	6.95 (2.86)
QUITA	HRISK	1.55 (0.42)

a. Each equation also includes the variables *AGE, BLACK, MALE, EDUC, MARRIED, KIDS, EXPER* (in *WPREM* equations), *TENURE* (in all except *WPREM* equations), and *EARNG* (in all except *WPREM* equations).

mium they received was sufficiently large to make the chemical industry salary substantially different from what could be earned elsewhere.

Some portion of this group who perceive risks but not relative risk premiums may, however, be mismatched. On a conceptual basis, there clearly is some potential for some labor market mismatches even with rational behavior if workers continually update their imperfect knowledge of the risks of the job as they acquire additional information through on-the-job experience (see Viscusi 1979 for a formal presentation of this model). Wage premiums for risk will be observed, but workers in high risk situations will also tend to quit once they have learned about the risks and have decided that the risk compensation is insufficient. Although past empirical work has focused on worker quitting,[4] a related prediction is that if workers were asked to repeat their job choice on the basis of current information, many workers in high-risk jobs would be reluctant to do so. Unlike worker quitting, this job acceptance question is not influenced by transactions costs of job changes, such as seniority rights. This question also avoids the limitations of the relative risk premium question, which may not fully capture the overall desirability of the job.

For the full sample, 79 percent of the sample would decide without hesitation to take the same job (TAKEA). The remaining 21 percent would either have some second thoughts or would definitely not take the job. Since 97 percent of all respondents would, at most, "have some second thoughts" (TAKEB), only 3 percent of the sample appears to have strong reservations about their positions. The combination of the wage premium estimates and the widespread willingness to repeat the employment decision suggests that job risks are not a major source of worker dissatisfaction. Few workers appear to be seriously mismatched.

One mechanism by which mismatches are remedied is through worker quitting. To analyze the job hazard–quit relationship, we developed quit intention variables utilizing the same phrasing as the Survey of Working Conditions questions analyzed in Viscusi (1979). As shown in that study, this quit intention measure yielded results quite similar to those generated by actual quit behavior. One-eighth of the sample was very likely or somewhat likely to "make a genuine effort to find a new job with another employer within the next year" (QUITA), but only 5 percent were very likely to do so (QUITB). Some worker dissatisfaction is clearly present, but there is not a large proportion of severely dissatisfied workers at the firms in our sample.

The worker's job risk plays an instrumental role in the cases in which mismatches are observed. Table 6.3 presents the maximum likelihood estimates for the determinants of two job satisfaction measures. In each case, the equations also included a series of variables, such as worker age, that are strongly linked to worker turnover. The probability that the worker would repeat this job decision (TAKEA) is negatively related to all perceived risk variables, controlling for worker earnings and other related factors. A worker who views his job as dangerous (DANGER), for example, will have a probability of repeating his initial job choice that is .22 lower than those who do not. Similarly, all of the job risk variables exert a positive influence on QUITA, where the quit intention probability will increase from .06 to .19—or over triple—if the worker views his job as dangerous. Put somewhat differently, the mean effect of the *DANGER* variable accounts for one-half of all quit intentions.

These results are consistent with a model in which the worker's job choice among potentially hazardous jobs is part of a continuing adaptive process. Workers' reservation wages will increase as their perceived risks rise, so that we will observe risk premiums for prior perceived risks and for some risks discovered on the job. Risks that workers learn about but for which they are not sufficiently compensated will generate quits. Although the evidence is consistent with this general view, the intermediate learning linkage and the behavioral implications of changes in risk assessments have not yet been examined.

6.2 Hazard Information and Risk Perceptions

Chemical Labels

To obtain evidence on workers' risk information processing, we carried out the following experiment in the second part of the questionnaire. We presented each worker with a hazard warning label for one chemical that was not a current part of his job. Each respondent was told that he would use 100-pound containers of this substance in his current job operations, but that this chemical would replace the chemicals with which he was currently working. The scenario was similar to that in which a worker learns that the chemicals he uses have been mislabeled. We provided workers with "new information" rather than informing them of existing hazards so as to be able to distinguish the role of the hazard warning from a priori knowledge about the job, thus providing a context in which learning could be observed. We then asked the workers how this change would affect their risk perceptions and other aspects of their behavior. As a result, subsequent changes in risk perceptions would not reflect an inadequacy in workers' prior judgments, but rather how information regarding a newly introduced risk would alter the assessment of the job's implications.

We assigned workers to one of four different labeling groups: sodium bicarbonate (CARB), a lachrymator chloroacetophenone (LAC), asbestos (ASB), and TNT. The CARB control group was kept relatively small because our primary focus was on the implications of the three risky substances. Each worker was given the information on a label with the standard format now used in the chemical industry. Figure 6.1 reproduces the labels exactly except that

CUNYCARB
SODIUM BICARBONATE

INDUSTRIAL

U.S.P./FOOD CHEMICALS CODEX GRADE

SPILL: Sweep-up, place in an appropriate chemical waste container. Flush area with water. Observe all local, state and federal regulations regarding disposal, spill, cleanup, removal or discharge.

100 LB. NET WT. 45.4 kg.

LOT NO.

CUNY CHEMICAL COMPANY
CHEMICAL SYSTEMS DIVISION — NEW YORK, NY 10033

CUNYLAC
CHLOROACETOPHENONE

WARNING! LACHRYMATOR - VAPOR AND DUST

EXTREMELY IRRITATING

•Do not breathe dust or vapor. •Wear a self-contained breathing apparatus, rubber gloves, and protective clothing when handling. •Use only with adequate ventilation. •Keep container closed. •Avoid contact with skin and eyes. •Wash thoroughly after handling.

FIRST AID: *If inhaled,* remove to fresh air. If not breathing give artificial respiration, preferably mouth-to-mouth. If breathing is difficult, give oxygen. Call a physician.

In case of contact, immediately flush eyes or skin with plenty of water for at least 15 minutes. Call a physician. Wash clothing before reuse. Discard contaminated shoes.

If swallowed, immediately dilute the swallowed material by rapidly giving large quantities of water and induce vomiting by gagging the victim with a finger or blunt object placed on the back of the victim's tongue. Continue fluid administration until vomitus is clear. Never give anything by mouth to an unconscious person. Call a physician or the nearest Poison Control Center immediately.

IN CASE OF FIRE: When heated emits toxic fumes. Wear self-contained breathing apparatus. Use water spray, foam, dry chemical or CO_2.

SPILL: Sweep-up, place in an appropriate chemical waste container. Flush area with water. Observe all local, state and federal regulations regarding disposal, spill, cleanup, removal or discharge.

KEEP OUT OF REACH OF CHILDREN   FOR INDUSTRIAL USE ONLY

UN 1697

IRRITANT

CHLOROACETOPHENONE, SOLID

100 LB. NET WT. 45.4 kg.

LOT NO.

CUNY CHEMICAL COMPANY
CHEMICAL SYSTEMS DIVISION — NEW YORK, NY 10033

Figure 6.1 Chemical labels for the job risk study. *Top*: CARB. *Bottom*: LAC.

ASBESTOP
ASBESTOS

DANGER!   CANCER HAZARD

Caution, Contains Asbestos Fibers • Avoid Creating Dust
• Breathing Asbestos Dust May Cause Serious Bodily Harm

• Use with a NIOSH/Mesa approved respirator • Use with approved goggles • Do not smoke in work area • Wash thoroughly after handling, before eating and before leaving work • Change clothing often; do not wash workclothes at home.

SPILL: Sweep-up, place in an appropriate, chemical waste container. Flush area with water. Observe all local, state and federal regulations regarding disposal, spill, cleanup, removal or discharge.

ORM - C

ASBESTOS

KEEP OUT OF REACH OF CHILDREN   FOR MANUFACTURING USE ONLY

100 LB. NET WT. 45.4 kg.

LOT NO.

Cuny
CHEMICALS

CUNY CHEMICAL COMPANY
CHEMICAL SYSTEMS DIVISION — NEW YORK, NY 10033

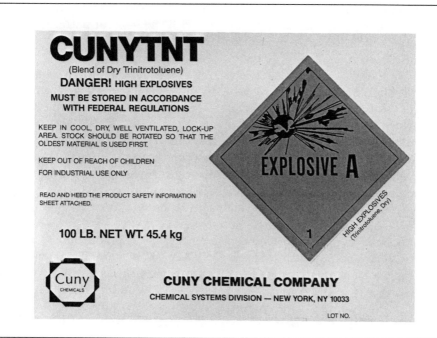

Figure 6.1 (cont.) *Top*: ASBESTOP. *Bottom*: TNT.

on the TNT label the diamond shape surrounding "Explosive A" was red. The CUNY prefix on the chemicals was derived from the acronym for the City University of New York, where O'Connor was a graduate student. All the labels were designed by O'Connor, who is a chemical industry labeling expert (see, for example, O'Connor and Lirtzman 1984).

Workers did not proceed with the rest of the questionnaire until they had answered successfully some basic general questions to test whether they had read the label. The workers appeared to have little difficulty in this regard, no doubt because they had substantial experience using chemicals labeled in this manner. Unlike the consumer study, we were not concerned with whether consumers read the labels in practice. Rather, assuming that workers do read the labels, what will be their response? Shortcomings in label comprehension can be ameliorated by improved label design—a matter that we excluded from consideration in the worker study.

Effects on the Key Variables

Although the information provided did not involve a specific risk level but rather a chemical hazard for which risk assessments will vary, the responses were consistent with the general patterns we expected. Table 6.4 summarizes the variable means for each labeling group.

The sodium bicarbonate label, for a very safe substance, led to a reduction in the *RISK* variable from .12 to .06 for *RISK1*, where the postscript 1 indicates the postinformation analogue of the variable. Besides halving the assessed *RISK* level, CARB also dramatically reduced the fraction of workers who believed they faced above-average risks. Only one respondent raised his *RISK* assessment (from .05 to .06), but since this worker was in a very low-risk job and had a posterior *RISK1* value identical to the CARB subsample mean, this behavior cannot be regarded as irrational.

If CARB were the only risk posed by the workers' jobs, one would expect them to assess this risk as being zero. Even for people working with a safe substance, however, there is a residual risk such as the risk of a safety-related job injury from accidents. Since the *RISK1* value of .06 for CARB equals the 1980 and 1981 average BLS injury rate for the chemical industry, the results are not out of line with what might be expected once the chemical hazards have been eliminated.

In addition, the assessed *RISK1* for CARB may not be zero because not all workers know what sodium bicarbonate is. The label

Table 6.4 Means of variables for each labeling group

Risk variable	CARB ($n = 31$)	LAC ($n = 106$)	ASB ($n = 102$)	TNT ($n = 96$)
RISK	.12	.10	.09	.10
RISK1	.06	.18	.26	.31
HRISK	.42	.38	.29	.40
HRISK1	.07	.83	.95	.98
WBOOST	.03	.48	.71	.82
Risk premium ($1982) ($Y1 - Y$)[a]	0	1,919.01	2,995.59	5,158.31
NOWAGE	.00	.02	.11	.17
QUITA	.23	.10	.13	.10
QUITA1	.00	.23	.65	.73
TAKEA	.67	.82	.80	.76
TAKEA1	.90	.58	.11	.07

a. The risk premium figures are conditional upon facing an increased risk and being willing to accept a finite risk premium.

suggests that it is a very safe chemical, but it does not explicitly state that it is risk free. In a series of seminar presentations by one of the authors to economists and lawyers, it became apparent that a substantial portion of the population was unaware that sodium bicarbonate and household baking soda were identical products. In many cases, individuals claimed that the mere presence of a label and the instructions to sweep up and dispose of sodium bicarbonate spills properly led them to believe that this chemical was risky.

The lachrymator was the second safest substance in the labeling group. Workers viewed this chemical as more hazardous than their present environment. The *RISK* level almost doubled, and the fraction of workers who considered themselves in above-average-risk jobs increased by .45. Eleven workers did not revise their risk assessments upward after seeing the LAC label, but these workers were in very high-risk jobs. Their *RISK* level decreased from .19 to .15, which is still above the average preinformation *RISK* value for the sample. Notwithstanding the absence of any assessed increase in risk for this subgroup, one person indicated that he was somewhat likely to look for a new job (*QUITA1*) even though he had not expressed this intention earlier. The low percentage of these inconsistencies is, however, reassuring.

The asbestos warning led to a more dramatic response. The riskiness of this substance relative to TNT is not clear-cut because of the deferred nature of asbestos-related cancers. Asbestos is, however, believed to be the cause of the majority of all occupational cancers. The asbestos label led workers to triple their assessed *RISK* levels, with almost all workers viewing their jobs as above average in riskiness.

Somewhat surprisingly, 5 percent of all workers did not view ASB jobs as posing above-average risk. Moreover, a substantial group of twenty-six workers, most of whom were in very high-risk jobs, did not raise their risk perceptions. This group's reservation wage and quit responses (that is, no increase in quits and elimination of all *QUITA* = 1 values) were consistent with their *RISK1* values; thus the *RISK1* variable appears to reflect a more favorable assessment of the job's attractiveness. Such a favorable response is not implausible, particularly for researchers who work with new unregulated carcinogens on a daily basis.

The explosive hazards of TNT generated the greatest risk assessment response, as all but two workers now viewed their jobs as above average in risk. Although eleven workers did not raise their *RISK* assessments in response to the warning, these workers were on very hazardous jobs (*RISK* equal to .19), and on average the TNT warning lowered their *RISK* value by only .04. One seemingly inconsistent respondent indicated that he was somewhat likely to quit (*QUITA1*) even though he hadn't been earlier and his assessed *RISK* level had not increased. As with the earlier results, the response to the information was overwhelmingly in the expected direction. The behavior of only a small minority of the workers does not appear consistent with a rational learning process. This result does not, however, imply that workers respond perfectly to new information, because the relation between the four labels and actual risk levels is not narrowly defined. Some imprecision is inherent because of differences in individual susceptibility to risk.

6.3 Formal Analysis of Worker Learning

Theoretical Model

To test the empirical implications of the hazard warnings more fully, we formalize the nature of the learning process. Our assumption here is that workers adopt a Bayesian learning approach in which we characterize their assessed probabilities as belonging to the beta family. This probability distribution is ideally suited to analyzing independent Bernoulli trials on events such as whether or not one suffers a job accident.[5] We view the receipt of the new labeling information as equivalent to observing additional Bernoulli trials concerning the riskiness of the job. Thus our implicit assumption is that labels simply serve to augment the risk information available to workers.[6] We do not rule out the possibility that this new information is regarded as perfect information.

The two parameters of the prior distribution are p, the prior assessed probability of an adverse outcome (that is, *RISK*), and γ, a term that can be regarded as the precision of the prior. After observing m unsuccessful outcomes (such as accidents) and n successful outcomes, the worker assesses the value of the posterior accident probability as being $(\gamma p + m)/(\gamma + m + n)$. The term γ is tantamount to the number of trials the worker acts as if he has experienced when forming his prior.

The informational content of each label, i, likewise depends on two parameters: ξ_i, the precision of the information (that is, the equivalent number of observations $m + n$ reflected in the information) and s_i, which is the fraction of these ξ_i observations that are unfavorable. Whether or not the label raises workers' probability assessments depends on whether s_i exceeds p, and the extent of revision is positively related to the informational content ξ_i, for any given value of s_i. If workers are provided with perfect information and if the labeled chemical is the only risk, the value of ξ_i should be infinite. The labels do not specify the exact chemical risk, so that ξ_i need not be infinite in practice. Moreover, the label conveys only information regarding the risks from direct chemical use, so that all accident-related risks and all environmental chemical risks remain. Worker responses consequently will reflect the relative weights workers placed on the prior and posterior information. These weights will capture both the degree to which the information in the label was credible and the relative role of this risk in the new version of the worker's job.

The posterior probability p_i of an adverse job outcome after receiving a hazard warning for chemical i is given by:

$$(6.1) \qquad p_i = \frac{\gamma p + \xi_i s_i}{\gamma + \xi_i} = \frac{\xi_i s_i}{\gamma + \xi_i} + \frac{\gamma p}{\gamma + \xi_i}.$$

The regression equation counterpart of equation 6.1 for each chemical i is

$$RISK1_i = \alpha_i + \beta_i RISK_i + u_i,$$

where u_i is a random error term, and

(6.2) $$\alpha_i = \frac{\xi_i s_i}{\gamma + \xi_i} \text{ and } \beta_i = \frac{\gamma}{\gamma + \xi_i}.$$

To take into account the bounded nature of the dependent variable, we also estimate the equations in terms of the log–odds of the probability, or $\ln[RISK/(1 - RISK)]$. In this case, the parameters α_i and β_i for the linear regression counterpart can be derived from the regression results but are not produced as directly.

The estimated versions of the parameters in equation 6.2 also can be used to construct two key measures of the information conveyed by the warning. The risk level s_i is given by

(6.3) $$s_i = \frac{-\alpha_i}{\beta_i - 1},$$

which can be verified using Equation 6.2. Similarly, the informational content of the warning relative to the prior, Ψ_i, is given by:

(6.4) $$\Psi_i \equiv \frac{\xi_i}{\gamma} = \frac{1}{\beta_i} - 1.$$

Higher values of Ψ_i imply greater informativeness of the label relative to the worker's initial judgments.

To the extent that workers' risk responses reflect not only changes in the probability of an adverse outcome but also changes in their severity, we must modify the formulas above. Let V_i be the severity (that is, monetary equivalent) of the health impact posed by the hazard warning relative to that posed by the average U.S. job injury, which serves as the metric for the analysis. If the $RISK1$ responses reflect changes in both the probability of an accident and its severity, equation 6.3 becomes

(6.5) $$s_i V_i = \frac{-\alpha_i}{\beta_i - 1},$$

and the formulation and interpretation of equation 6.4 remain unaltered.[7] Although the discussion below is in probabilistic terms and does not include V_i explicitly, it should be noted that these risks are severity weighted.

Table 6.5 Risk perceptions after information: Regression results

Form of information provided		Constant		RISK		R²	s_i	Ψ_i
CARB	RISK1 (linear)	0.03	(0.01)	0.21	(0.10)	0.12	0.038	3.72
	RISK1 (log-odds)	−3.58	(0.46)	3.23	(3.56)	0.03	0.042	4.98
LAC	RISK1 (linear)	0.14	(0.01)	0.40	(0.10)	0.14	0.239	1.29
	RISK1 (log-odds)	−2.05	(0.13)	3.76	(1.12)	0.10	0.274	0.83
ASB	RISK1 (linear)	0.25	(0.02)	0.14	(0.14)	0.08	0.289	6.43
	RISK1 (log-odds)	−1.23	(0.11)	1.36	(1.03)	0.02	0.325	2.80
TNT	RISK1 (linear)	0.31	(0.02)	0.03	(0.13)	0.01	0.317	31.36
	RISK1 (log-odds)	−0.86	(0.09)	0.11	(0.76)	0.01	0.315	40.67

Empirical Results

Table 6.5 summarizes the regression results and the parameters calculated from them. Overall, the linear variant of the equation provided a better fit than the log-odds formulation. The coefficients α_i and β_i reflect the nature of the learning process. Figure 6.2 illustrates the range of possibilities. In the case in which workers' judgments are not affected by the hazard warning and are solely dependent on their prior value of *RISK*, α_i will equal 0 and β_i will equal 1. This no-learning case is the 45-degree line in Figure 6.2. At the other extreme, in which the hazard information is dominant, β_i will equal 0 and α_i will be positive. This case is the flat line in Figure 6.2. The regression results were between these two extremes. This intermediate learning case is the third possibility appearing

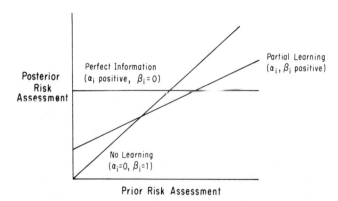

Figure 6.2 Learning and the empirical procedure.

in Figure 6.2. In all cases the label provided a substantial input, and in two cases the prior continued to play a significant role. These results are broadly consistent with a Bayesian learning model.

In the case of CARB, the label lowered the *RISK* assessment but did not eliminate the role of the prior: both α_i and β_i were statistically significant in the linear case in which the relation to equation 6.2 is direct. The risk level s_i implied by CARB was .04, or under half of the worker's prior *RISK* level, and Ψ_i implies that the relative precision of the hazard warning was four to five times that of the prior.

Since the very safe properties of sodium bicarbonate are reasonably well known, CARB might have been expected to result in a larger relative precision estimate and a lower s_i than was observed. A possible explanation is that workers did not place an infinite weight on a chemical exposure with near-zero risk because of the residual risks of the job. These workers will continue to be exposed to a variety of airborne carcinogens and safety-related risks that will be reflected in the posterior *RISK1* values. As the risks captured by the label approach zero, the nonzero risk components of the worker's job become more instrumental since they dominate the role of the label.

The CARB label was, however, much more powerful than the LAC warning. This label led to the greatest retention of workers' prior beliefs: the *RISK* coefficients are the largest of any of the regressions. A small impact was not a consequence of any close similarity in the hazard probabilities of LAC and *RISK*, since s_2 is over double the prior value of *RISK*. The limited nature of the effect derives from the lower relative precision, Ψ_2, of this warning, which had roughly the same informational content as workers' prior beliefs.

Warnings for the severe risks of ASB and TNT are so powerful that the prior *RISK* variable plays an insignificant role; only the constant terms enter. The risk levels s_3 and s_4 are somewhat higher than for LAC, but the major difference is the precision of the information. Asbestos warnings have roughly the same relative precision as LAC, but TNT has especially large informational content, roughly thirty to forty times that of the prior. Since TNT poses well-known explosive risks, this result is not unexpected.

Overall, the risk levels s_i implied by LAC, ASB, and TNT were not too dissimilar. The greatest difference was the relative precision associated with these warnings. The impact of a hazard warning does not hinge solely on the implied risk level. In this instance, the informational content of the label proved to be more instrumental in altering workers' probabilistic judgments. To be effective, hazard warnings must convey information in a convincing manner. Otherwise, the weight individuals place on their prior beliefs will dominate in the formation of workers' risk judgments.

6.4 The Effect of Learning on Worker Behavior

Overview

The change in the risk perceptions resulting from the hazard warnings will in turn affect worker behavior if workers make sequential decisions in an optimal manner. The data in the bottom half of Table 6.4 summarize the wage and turnover effects, which reflect similar patterns of influence.

The demand for risk premiums is positively related to the change in the risk, as we would expect. For ASB and TNT, about three-quarters of workers indicated they would require a higher salary to be willing to work with the new chemical (WBOOST). As noted above, some workers with very high initial risk assessments did not increase those assessments as a result of the label. The amount of extra compensation demanded ranges from $2,000 for LAC to over $5,000 for TNT. Workers needed no risk premium to work with CARB. (Indeed, they should be willing to take a pay cut, but the survey did not address this possibility.) The premium estimates are only for workers willing to remain on the job in return for extra pay. Some workers, particularly for TNT and ASB, were not willing to state an acceptable reservation wage (*NOWAGE*). Whether these twenty-nine nonrespondents were unwilling to accept any finite risk premium or simply believed that no adequate risk premium was feasible is unclear.

The effect on worker turnover was particularly dramatic, since the experiment altered the risk but not the wage rate. Thus these risks produced a more dramatic worker response than in a market context, where there would be some adjustment in the wage level. The *QUITA1* and *TAKEA1* questions pertained to the attractiveness of the current job. Among workers given the CARB label, quit intentions dropped from 23 percent to zero, with a corresponding increase in the percentage of workers who would repeat their job choice. The lachrymator produced a 13 percent increase in quit intentions and a 24 percent drop in willingness to repeat the job choice. As we would expect, the strongest effects were for ASB and TNT.

Regression Results

An instructive check on the validity of these responses is to analyze whether the behavioral relationship governing the risk premium and quit decisions parallel those in the preinformation situation. Such an analysis also makes possible an explicit test of the impact of the risk s_i implied by the label and its relative precision Ψ_i. Higher implied risks s_i clearly should make the job less attractive. The relative precision of workers' risk assessments will also increase workers' reservation wages, since, as shown in Viscusi (1979), the value of a risky job is negatively related to the precision of one's risk judgments. Jobs associated with looser probabilistic judgments are more attractive because they offer greater potential gains from experimentation. Workers can terminate uncertain jobs if their learning is unfavorable and reap the high expected rewards from jobs associated with favorable on-the-job experiences. This asymmetry generates a predilection for loose priors. This aspect of adaptive behavior is the most distinctive prediction of the model, but it has never been the subject of an explicit empirical test.

To analyze the effect of the hazard warnings on the level of compensating differentials, we let Y represent initial worker income, X a vector of all nonrisk variables for that job, Z the unmeasured effects specific to the job-worker match, and u the error term. The baseline compensating differential results in Section 6.1 focused on an equation of the form:

$$(6.6) \qquad Y = \beta X + \beta^* RISK + \beta^{**}\gamma + Z + u.$$

Since γ and Z were omitted from the model, the estimated coefficients were subject to omitted variables bias.

The situation following information (denoted by postscript 1) can be modeled similarly, where:

$$(6.7) \qquad Y1 = \beta X + \beta^* RISK1 + \beta^{**}(\gamma+\xi) + Z + u1.$$

Subtracting equation 6.6 from equation 6.7 yields:

$$(6.8) \qquad (Y1 - Y) = \beta^* \Delta RISK + \beta^{**}\xi + (u1 - u),$$

where $\Delta RISK$ is $RISK1 - RISK$. Equation 6.8 will yield consistent estimates of the coefficients in this fixed effects model as the sample size $N \longrightarrow \infty$ if there is sufficient variation in $\Delta RISK$ and ξ.[8] We do not have information on ξ_i but rather on Ψ_i for each labeling group, which is ξ_i/γ. Workers, however, will differ in the precision of their priors, so that γ will be a random variable. Since the workers were

Table 6.6 Postinformation earnings and quit equations[a]

Dependent variable	Risk or ΔRISK	s	Ψ	$R^2/-2$ log likelihood
EARNG	9,934.5	6,784.2[b]	52.0[b]	0.24
	(5,468.6)	(3,342.3)	(32.2)	
ΔEARNG	12,435.3	—	65.56[b]	0.17
	(2,681.9)	—	(20.10)	
EARNG	9,838.3	6,602.3[c]	41.6[c]	0.24
	(5,471.3)	(2,684.2)	(19.6)	
ΔEARNG	12,777.5	—	46.53[c]	0.17
	(2,640.3)	—	(14.17)	
LNEARNG	0.627	0.456[b]	0.0021[b]	0.28
	(0.303)	(0.185)	(0.0007)	
ΔLNEARNG	0.633	—	0.0027[b]	0.31
	(0.087)	—	(0.0007)	
LNEARNG	0.622	0.424[c]	0.0018[c]	0.28
	(0.303)	(0.149)	(0.0011)	
ΔLNEARNG	0.651	—	0.0019[c]	0.31
	(0.086)	—	(0.0005)	
QUITA	−1.05	5.95[b]	0.027[b]	381.30
	(2.03)	(1.43)	(0.013)	
ΔQUITA	20.4	—	0.011[b]	63.30
	(4.3)	—	(0.029)	
QUITA	−1.07	5.75[c]	0.021[c]	384.60
	(2.02)	(1.17)	(0.008)	
ΔQUITA	20.6	—	0.002[c]	63.50
	(4.3)	—	(0.020)	

a. All cross-sectional equations include other explanatory variables as in Tables 6.2 and 6.3.
b. The Ψ variable is based on the linear regression estimates reported in Table 6.5.
c. The Ψ variable is based on the log-odds regression estimates reported in Table 6.5.

assigned randomly to each labeling group, the precision variable should be subject to random measurement error, biasing the β^{**} coefficient downward.

Table 6.6 reports the earnings equations both in the first difference form (that is, ΔEARNG, ΔLNEARNG) and in the cross-sectional form for the postinformation case, where RISK is of the same form as the dependent variable (ΔRISK). Because the first differencing eliminates the biases from omitted fixed effects, the change-in-earnings equations are estimated for the full sample, while the cross-sectional results focus on the BC/TECH subsample as before. For the postinformation cross-section, we included both RISK and *s* rather than *RISK1* in order to estimate explicitly the role of the risk implied by the warning.The results reflect a consistent pattern of premiums for prior risks and risk communicated through the label. Similarly, labels associated with high relative precision Ψ generate additional premiums, as predicted.

The consistency of worker behavior with the earlier results is more difficult to ascertain. Premiums per unit of risk should be larger because individuals will demand higher rates of compensation if placed in a highly risky job that is not consistent with their preferences. Whereas the initially perceived risks are the result of a voluntary self-selection process, the postinformation risks are not, and serious mismatches may occur. Consequently, higher desired premiums per unit of risk should be observed.

The magnitude of the postinformation wage-risk tradeoff bears out this pattern. In the case of the linear specifications, for example, the *RISK* and *s* coefficients average about one-fifth higher than in Table 6.2, whereas in the first difference form Δ*RISK* commands premiums three-fourths larger. A greater response is observed in the first differencing case because the additional desired premiums per unit of risk for the added hazard will be averaged only across the extra risks, whereas the postinformation cross-section obtains an average unit risk premium for the entire risk level. In addition, about one-third of the discrepancy arises because the first differencing results focus on the full sample, which is wealthier than the BC/TECH subsample used in the cross-sectional results. These workers consequently demand a larger premium per unit risk.[9]

To analyze the change in workers' quit decisions, we can formulate a postinformation cross-section and an analogue of the fixed effects model for discrete variables.[10] The postinformation quit intentions in the cross-sectional results are driven exclusively by the implied risk and precision of the hazard, each of which has the expected positive effect. The most dramatic difference with the earlier results is in the Δ*RISK* coefficients in the first difference equations, which are almost three times larger than in the preinformation results in Table 6.3. Such a dramatic increase is not implausible, since quits arising in the market are in response to a pay-risk package mix that the worker initially accepted. Here workers are responding to often dramatic changes in their job's attractiveness so that the intensity of the response should increase. The Δ*QUITA* equations do not, however, lead to significant coefficients for Ψ. This result may be due to the drop in sample size to 161 as a consequence of the statistical estimation procedure used.

6.5 Conclusion

This chapter has focused on an adaptive framework in which individuals do not have perfect job risk information, but instead continually revise their risk judgments in Bayesian fashion and then switch jobs once these judgments become too unfavorable. This theory is an extension of the standard compensating differential analysis rather than an incompatible theory. Workers' initial perceptions of risk led to compensating differentials and also generated intentions to quit and regret over having accepted the job initially. The evidence of risk-related job mismatches is consistent with a model of job experimentation and would not occur in a perfect information version of the compensating differential model. The extent of these mismatches does not, however, appear to be great; for this sample the market appears to operate reasonably effectively.

After being given a hazard warning label for a new chemical to be used in their job, workers revised their risk assessments in the expected directions. The new information was not always dominant, however; workers retained some influence of their prior risk beliefs, particularly for hazard warnings with low informational content.

A particularly important distinction affecting the impact of the different labels is the informational content of the warning, not simply the risk level implied. In particular, what relative weight do workers give the new information when updating their risk perceptions? Increases in the informational content boost the effectiveness of a hazard warning for any particular level of risk.

Although the risk level implied by the label was of consequence, differences in informational content Ψ_i appeared to be more influential in governing workers' posterior risk assessments. This learning in turn generated a demand for risk premiums and incentives to quit, as expected. Both the change in the level of the risk and changes in the precision of workers' judgments were of consequence, as predicted by an optimizing model of workers' experimentation process with uncertain jobs. Although the change in the risk level had a more consistent direct effect on behavior than did the relative precision of the hazard warning, the precision also had an indirect influence through its powerful impact on the posterior risk assessment.

The pivotal influence of the informational content of the chemical label has broad ramifications for the design of effective risk information strategies. Past informational campaigns such as those intended to encourage seatbelt use and deter cigarette smoking have had disappointing results. The primary purpose of these efforts has been to exhort rather than to provide consumers with information that they did not already possess. The lack of a major consumer response is not surprising, because the informational content of these warnings has been low. The results in this chapter indicate that to be most effective, risk information programs must not simply convey the risk level but must also provide individuals with new information in a convincing manner.

Most workers behaved as expected, but there was a small minority of alarmist responses as well as some inertia and inconsistencies. Moreover, even though the empirical evidence constitutes the most refined test of the Bayesian learning model of adaptive job choice, observed consistency with the principal predictions of the theory does not necessarily imply full rationality. Since the Bayesian learning model represents an intermediate case, falling between the no learning situation and the perfect information case, a fully conclusive test of consistency is more difficult than in many statistical contexts. Nevertheless, there is strong evidence of a systematic worker response that is quite different from the polar extremes of optimal decisions with perfect prior information and random decisions by irrational workers.

Although the principal objective of the worker labeling experiment was to develop new results with respect to learning, an additional dividend was the parallelism between some of our findings and earlier results. In a few cases it was possible to make direct comparisons of the response to the hypothetical labeling treatments and workers' current behavior with respect to actual job risks. The strong similarity in the wage and quit equations with those in the literature bolsters the plausibility of the survey results for responses to chemical labels. A likely contributor to the success of the labeling experiment is the strong linkage between the experiment and an actual work situation in which labels such as these could be used.

7 | Implications for Economic Behavior

WESLEY A. MAGAT and
W. KIP VISCUSI

The fundamental premise of this book is that information policies affect economic behavior by altering decisionmakers' prior beliefs about risks. Changes in their perceptions of risk as a result of the information program cause them to revise their estimates of the risks associated with their actions, which in turn lead them to select a different level of precautionary behavior or to demand a higher reward for engaging in the risky activity. This Bayesian paradigm of individual responses to risk information policies guided our hypothesis testing about the determinants of precautionary behavior in using consumer products and of wage risk premiums for working in hazardous workplaces.

The general issue we addressed was how individuals process information about risks and how this learning affects subsequent behavior. Our findings confirm that individuals do respond to information policies in ways that are generally consistent with the Bayesian learning model. In addition, decisions about precaution-taking appear to be reasonably consistent with individuals' preferences for the benefits of injury avoidance and for the disutility costs of precautionary actions.

These findings provide support for the efficacy of information policies to control the hazards from dangerous products. Not all individuals will behave optimally, however, because such decisions under uncertainty involve inherent shortcomings. Perhaps the chief limitation is that if individuals have strongly held initial beliefs about the product, then informational programs must be very persuasive if they are to alter these beliefs. Information policies should be designed both to convey accurately the best objective estimates of product risks and to convince users of the accuracy of those estimates.

The awareness of the imperfections of the information provision approach to controlling hazards does not necessarily imply that banning products or restricting their uses provides a superior control approach. Product use restrictions, even if perfectly enforced (which is both unlikely and costly), suffer from two flaws that are quite different from the shortcomings of the information provision approach. They do not allow variations in precautionary behavior across product users who may differ in their intended uses, their volume of product use, their aversion to taking precautions, and, most important, their preferences for avoiding the product hazards. In addition, to make informed decisions about product use restrictions, policymakers must estimate individual preferences for both hazard avoidance and the disutility of taking precautionary actions. Information provision avoids both of these problems. Thus, deciding about which policy approach to take toward risky products requires comparing the magnitudes of the imperfections of both approaches.

Our findings provide important evidence of the extent to which information policies induce product users to select the level of precautionary behavior consistent with their own preferences and the best objective estimates of risks. Similarly, we showed that more fundamental decisions about whether to engage in risky activities, such as workers' job choices, are also quite responsive to the risk information provided.

To guide the hypotheses being tested, we developed an economic model of how consumers' precautionary effort responds to various characteristics of the information provided by product labels. We based the design of these labels on the principles for effective information provision that have been developed in the marketing and psychology literature. Chief among these concerns is the need to provide individuals with information about the risks, the appropriate actions, and, consequently, the benefits of these actions, in a form that can be readily processed. Consumers did respond to the labels in a manner that was broadly consistent with the predictions of the economic model.

Large risks led consumers to take more precautionary actions than did small risks, as expected. More important, the greater the amount of risk information on the labels, the more extensive the precautionary responses to them. This finding held both for the relationship between the total amount of risk information on the labels and the several measures of precautionary behavior, and between information about specific risks, such as chloramine gassing, and the specific precautions necessary to avoid them.

As predicted in Chapter 2, the format of the label designs was an important determinant of their effectiveness in causing consumers to take precautions. Adjusting for other factors that influence effectiveness, such as demographic characteristics, the labels designed in accordance with the principles discussed in Chapter 2 generally led to more precaution-taking than did labels not formatted to take into account the cognitive processes that determine how people use labeling information.

By analyzing the implicit values of avoiding the injuries associated with each of the precautionary actions, we discovered some potential shortcomings in informational strategies or in consumer responses to them. In particular, it appears that consumers underreacted to the risks associated with storing drain opener in locations accessible to children. This result implies either that precautionary behavior is not fully rational or that the provision of information regarding this particular risk should be improved.

In contrast to the findings above, which suggest that consumers' precaution-taking responses are generally consistent with the patterns of behavior predicted by our economic model, Chapter 5 found that consumers tend to attach overly high dollar values to the avoidance of the nonfatal injuries from consumer products relative to health valuations derived from market-based behavior. This high rate of tradeoff between dollars and risks is consistent with much experimental evidence on individuals' reactions to low probability events, but it is inconsistent with observed failures of individuals to purchase subsidized insurance against natural hazards such as floods, which are also low-probability events.

A possible explanation for these conflicting results is that the extent to which consumers are *forced* to make decisions involving low-probability events is the crucial element of the decision context that determines whether they attach a high or a low value to the risks. If they are not forced to make a decision, such as the purchase of a risky product or insurance against some risks, then they may ignore the risk by treating the probability as zero. However, when a risk information program forces them to focus on the probability that some rare injury will occur, they may treat the probability as if it were higher than its objective value because of the cognitive difficulties in making risky choices when the probabilities are extremely low. This behavior leads to the imputation of a high value for the avoidance of the injury.

This finding identifies the competing concerns in the design of labels. If the risks are not emphasized on the labels, users may ignore them. Yet if the risks are given too much prominence relative to the information about the efficacious use of the products, consumers may overreact to the risks and avoid the use of the product, substituting other products and actions that may be less useful and potentially more risky.

Chapter 5 also explored the accuracy of consumer risk perceptions. When assessed in terms of their family's risk of injury *relative* to the average household using the product, almost all consumers considered their household to be as safe or safer than the average household, an obvious inconsistency. In contrast, Chapter 6 measured the risk perceptions of workers in chemical plants for the risks they face on the job and found risk perceptions close to the levels commonly found in the chemical industry. It is possible that individuals assess the levels of the the risks they face fairly accurately but tend to be overly optimistic in assessing the relative safety of their own behavior. Thus, the framing of the survey questions may be instrumental. Alternatively, the risk perceptions about job risks may be much more accurate than those associated with the use of hazardous products in the home, which involve much lower-probability events for which the role of individual precautions is much greater.

Chapter 6 explored how workers respond to information about risks on the labels of hazardous chemicals, tracking the linkage from labeling information about risks and precautions to changes in risk perceptions, to subsequent changes in labor market behavior, such as demanding additional wages for risky jobs and quitting. By monitoring changes in risk perceptions resulting from the receipt of new information on labels, this study was able to demonstrate that workers revise their risk assessments in the expected directions and that the change in these perceptions is instrumental in determining worker behavior.

A major focus of our worker risk study was not only the role of the risk conveyed by the label, but also the role of its informational content. The strength of the message conveyed was a primary determinant of the degree to which workers' risk perceptions were revised, with high information content needed to make large revisions in risk perceptions.

The findings also show a strong link between the risk information on the labels and worker responses to it. Specifically, higher risks led to both a demand for more risk compensation and higher intended quit rates. The precision of the risk information conveyed by product labels also affected the behavioral responses of workers as predicted by a model of optimal worker experimentation with hazardous jobs. The strength of this effect was smaller than that of the risk level itself, but the degree of information provided influenced behavior indirectly through its strong effect on the posterior risk assessments.

The important role played by informational content on job risk labels highlights a fundamental feature of informational efforts. To affect individual learning about risks, these policies must provide new information in a convincing manner. Policies of education that simply attempt to persuade consumers to alter their behavior through repetition of information that is already well known will be less effective.

The credibility of the findings in this chapter is reinforced by the close parallels between the wage and quit equations estimated from survey responses about the workers' current jobs and similar equations for actual worker behavior.

Although the surveys of consumer and worker behavior addressed many aspects of the role of risk information, much remains to be learned about informational policies, particularly about the nature of individual behavior and the practice of label design.

Many of the doubts concerning the efficacy of informational regulations stem from instances of irrational behavior observed in risk studies by psychologists. For example, to what extent do individuals overreact to low-probability events, and how does the context of the risk information influence this type of behavior, as opposed to the opposite problem of ignoring the risks? It would be interesting to assess the degree to which individuals' implicit valuations of health outcomes and their risk assessments are sensitive to the level of the risk involved, the type of risk (for example, immediate versus long-term carcinogenic), the individual's ability to influence the risk through appropriate precautions, and the type of activity involved (such as household cleaning or agricultural pesticide use).

Our knowledge of effective label design for risky products remains quite rudimentary. Little is known, for example, about which symbols, warning language, and organization are most effective. One possible approach to these uncertainties is to undertake actual field tests of labels, much like the surveys reported here, to ascertain the most effective label design for a particular context.

Overall, the results reported here suggest that information policies are a promising approach to controlling risks. Properly designed information provision programs have the potential to improve significantly the quality of decisions that consumers and workers make about the use of hazardous products. However, until we know more about how effective information programs can be in different situations, it will be difficult to evaluate when risk problems can be more effectively controlled by information programs or by more direct regulatory interventions.

Appendixes

Notes

References

Index

APPENDIX A

Cleaning Agent Questionnaire

SUBJECT # _____

INTERVIEWER NAME _____

Introduce yourself and ask subject if he/she is ready to begin. Go to Stage 1.

Stage 1: Examination of the Cleaning Agent

(Show subject the cleaning agent.)

Please examine this new cleaning agent as if you are about to use it in your home for the first time. I will ask you questions about the way you would use it. There are no right or wrong answers to any of the questions, just answer them as honestly as you can.

Take as much time as you would if you were going to use it in your home.

Stage 2: Questions about Usage

I will list several possible uses for this product. Please tell me if it is likely that you would use it in these ways:

Possible Use

2.1 For cleaning dirty sinks	yes	no	n/a
2.2 For removing mildew from walls	yes	no	n/a
2.3 For cleaning floors	yes	no	n/a
2.4 To add to toilet bowl cleaner if toilet is badly stained	yes	no	n/a
2.5 To add to wash for problem stains	yes	no	n/a
2.6 To add to ammonia or other ammonia-based cleaners for particularly dirty jobs	yes	no	n/a

Stage 3: Storage of the Product

3.1 Where would you store a product like this in your home?

(Do not read. Probe if necessary.
   In a childproof location? _____ Yes _____ No)

(Acceptable probes:
   Is that above or below your washing machine?
   Is that above or below the counter?)

Stage 4: Precautions Valuation--Cleaning Agent

Cleaning agents can be made a number of different ways. Presently, most products on the market have some potential drawbacks as well as benefits. For example, many cleaning agents have a bitter smell. To help us know which are most valuable to you, I am going to name a number of possible improvements in a 79 cent bottle of cleaning agent. For each tell me how much more, if anything, you would be willing to pay for the new product.

Current product	New product	What is the most you would pay OVER 79 cents for the new product?
4.1 Bitter smell	Fresh piney scent	_____ cents
4.2 Container cannot be reused	Reusable container	_____ cents
4.3 Cap cannot be used as a measure	Cap measures correct amount to use	_____ cents
4.4 Must be stored on a high shelf	Can be stored anywhere	_____ cents
4.5 Won't work if mixed with other cleaners	Effective if mixed with other cleaners	_____ cents

Stage 5: Frequency of Use Questions

5.1 How many bottles of bleach of the size we showed you does your household use in a given year?

(ask by month if necessary and then convert)
(bottles of bleach/year _____?)

Stage 6: Introduction to the Personal Computer

    Now let's turn to the personal computer and let you respond to questions from it. Each time a question appears on the screen, please read it and then answer the question. The computer will ask you to press a number on the keyboard that corresponds to your answer. (Show subject numeric keyboard.) Then you press the return key so the computer knows that you have answered the questions.
(Show subject the return key.)

    (If first question does not show up on the screen, insert disk and turn on the computer. The screen will show a question asking the subject to input his subject number. Tell subject his subject number and let him type it into the computer.)

    (Stay with the subject until he correctly answers the first question asking "What month is it?".)

    I will leave you to answer the remaining questions yourself. If you have questions, there is someone here to help you. (Point out the computer reference person.) Just raise your hand and she will be right over. When you finish, tell me and I will answer any questions you might have. (Thank subject and leave him at personal computer.)

BEFORE WE BEGIN, COULD YOU ENTER YOUR SUBJECT NUMBER (THIS IS A NUMBER GIVEN TO YOU BY THE PROGRAM OPERATOR).

AFTER TYPING IN YOUR NUMBER, PLEASE HIT THE RETURN KEY.

THANK YOU!

Hello! This is your first question on the computer.

What month is it?

[1] May   [2] June   [3] July

PRESS A NUMBER (1, 2, OR 3) INDICATING THE MONTH.

THEN PRESS THE RETURN KEY.

Well done! You are correct. It is June.

PRESS ANY KEY TO CONTINUE.

The cleaning agent you have been considering is a type of household bleach. To help you make decisions in answering the following questions, please think of the product as a bleach.

PRESS ANY KEY TO CONTINUE.

When bleach is mixed with ammonia or acid-based products like toilet bowl cleaners, chlorine gas forms. Breathing this gas causes headaches and burning lungs, eyes, and nose. The victim may need to be hospitalized for several days and recovers completely within a week.

PRESS ANY KEY TO CONTINUE.

Every year, this accident occurs to 50 out of 2,000,000 households (the number of households in North Carolina) which use bleach. Of course, this accident is more or less likely to happen in different homes depending on how bleach is used.

Suppose you used this product in YOUR home. Would the chance of an accidental gas poisoning be more or less likely in your home than in an average home that uses bleach?

[1] Less likely   [2] More likely   [3] About the same

PRESS NUMBER (1, 2, OR 3) ON YOUR KEYBOARD.

This means that you think an accidental gas poisoning from mixing bleach with another cleaner is more likely in an average home than in your home.

Indicate how much more likely this accident is in an average home that uses bleach than in your home.

In an AVERAGE home, a gas poisoning from mixing bleach with another cleaner would be:

    [1]   about as likely as in YOUR home
    [2]   twice as likely
    [4]   four times as likely
    [10]   ten times as likely
    [100]   one hundred times as likely

PLEASE TYPE IN A NUMBER INDICATING YOUR ANSWER. SELECT ONE OF THE FIVE POSSIBLE CHOICES OR ANOTHER NUMBER THAT BEST ANSWERS THE QUESTION.

The following questions will ask you to compare two bleach products. The CURRENT BLEACH, now sold in stores, is always listed on the left side of your screen. The NEW BLEACH is always found on the right side of your screen.

Let's try a practice question first.
PRESS ANY KEY TO CONTINUE.

```
            CURRENT BLEACH              NEW BLEACH
    * * * * * * * * * * *     * * * * * * * * * * *
    *                   *     *                   *
    * Cost per year:    *     * Cost per year:    *
    *    $10.00         *     *        ?          *
    *                   *     *                   *
    * Injury level:     *     * Injury level:     *
    *  50 gas poisonings *    *  50% DECREASE in  *
    *  for every        *     *  gas poisonings   *
    *  2,000,000 homes  *     *  compared to the  *
    *                   *     *  current product  *
    * * * * * * * * * * *     * * * * * * * * * * *
```

Think about how much you would be willing to pay for the bleach on the right.

Then . . . PRESS RETURN KEY TO CONTINUE.

```
            CURRENT BLEACH              NEW BLEACH
    * * * * * * * * * * *     * * * * * * * * * * *
    *                   *     *                   *
    * Cost per year:    *     * Cost per year:    *
    *    $10.00         *     *        ?          *
    *                   *     *                   *
    * Injury level:     *     * Injury level:     *
    *  50 gas poisonings *    *  50% DECREASE in  *
    *  for every        *     *  gas poisonings   *
    *  2,000,000 homes  *     *  compared to the  *
    *                   *     *  current product  *
    * * * * * * * * * * *     * * * * * * * * * * *
```

Let's make certain that you understand the question completely. The two bleach products are IDENTICAL IN ALL RESPECTS except for their chances of causing gas poisonings. The NEW BLEACH is 50% less likely to result in hand burnings in your home than the CURRENT BLEACH.

The amount you are willing to pay will affect only YOUR household's costs and not those of any other bleach users.

READY TO ANSWER THE QUESTION? PRESS ANY KEY TO CONTINUE.

```
         CURRENT BLEACH                   NEW BLEACH
* * * * * * * * * * * *          * * * * * * * * * * * *
*                      *          *                      *
* Cost per year:       *          * Cost per year:       *
*    $10.00            *          *       ?              *
*                      *          *                      *
* Injury level:        *          * Injury level:        *
*  50 gas poisonings   *          *  50% DECREASE in     *
*  for every           *          *  gas poisonings      *
*  2,000,000 homes     *          *  compared to the     *
*                      *          *  current product     *
* * * * * * * * * * * *          * * * * * * * * * * * *
```

Think about how much you would be willing to pay for the bleach on the right.

Would you be willing to pay more than $10.00 for the NEW BLEACH?

PRESS NUMBER (1) FOR YES OR NUMBER (2) FOR NO.

```
         CURRENT BLEACH                   NEW BLEACH
* * * * * * * * * * * *          * * * * * * * * * * * *
*                      *          *                      *
* Cost per year:       *          * Cost per year:       *
*    $10.00            *          *       ?              *
*                      *          *                      *
* Injury level:        *          * Injury level:        *
*  50 gas poisonings   *          *  50% DECREASE in     *
*  for every           *          *  gas poisonings      *
*  2,000,000 homes     *          *  compared to the     *
*                      *          *  current product     *
* * * * * * * * * * * *          * * * * * * * * * * * *
```

How high would the price of the NEW BLEACH have to be before you would rather buy the CURRENT BLEACH?

$ _____/year

PRESS THE NUMBERS THAT SHOW HOW MANY DOLLARS AND CENTS YOU ARE WILLING TO PAY.

Thank you . . .

We now want to ask you how much you prefer one bleach product over another.

PRESS ANY KEY TO CONTINUE.

```
            CURRENT BLEACH                  NEW BLEACH
     * * * * * * * * * * *        * * * * * * * * * * *
     *                   *        *                   *
     * Cost per year:    *        * Cost per year:    *
     *    $10.00         *        *    $15.00         *
     *                   *        *                   *
     * Injury level:     *        * Injury level:     *
     *  50 gas poisonings*        *  40% decrease in  *
     *  for every        *        *  gas poisonings   *
     *  2,000,000 homes  *        *  compared to the  *
     *                   *        *  current product  *
     * * * * * * * * * * *        * * * * * * * * * * *

                 Which bleach would YOU prefer?
Strongly                                            Strongly
Prefer                         Equal                Prefer
CURRENT:-----:-----:-----:-----:--x--:-----:-----:-----:----- : NEW
BLEACH    1     2     3     4     5     6     7     8     9    BLEACH
```

PRESS A NUMBER INDICATING YOUR PREFERENCE AND THEN PRESS THE RETURN KEY.

```
            CURRENT BLEACH                  NEW BLEACH
     * * * * * * * * * * *        * * * * * * * * * * *
     *                   *        *                   *
     * Cost per year:    *        * Cost per year:    *
     *    $10.00         *        *    $15.00         *
     *                   *        *                   *
     * Injury level:     *        * Injury level:     *
     *  50 gas poisonings*        *  40% decrease in  *
     *  for every        *        *  gas poisonings   *
     *  2,000,000 homes  *        *  compared to the  *
     *                   *        *  current product  *
     * * * * * * * * * * *        * * * * * * * * * * *

                 Which bleach would YOU prefer?
Strongly                                            Strongly
Prefer                         Equal                Prefer
CURRENT:-----:-----:-----:-----:--x--:-----:-----:-----:----- : NEW
BLEACH    1     2     3     4     5     6     7     8     9    BLEACH
```

See where the X is located. Does X represent your correct answer?

PRESS NUMBER (1) FOR YES, PRESS NUMBER (2) TO GET QUESTION AGAIN.

```
         CURRENT BLEACH                    NEW BLEACH
  * * * * * * * * * * * *         * * * * * * * * * * * *
  *                      *        *                      *
  * Cost per year:       *        * Cost per year:       *
  *    $10.00            *        *      $19.50          *
  *                      *        *                      *
  * Injury level:        *        * Injury level:        *
  *  50 gas poisonings   *        *  40% decrease in     *
  *  for every           *        *  gas poisonings      *
  *  2,000,000 homes     *        *  compared to the     *
  *                      *        *  current product     *
  * * * * * * * * * * * *         * * * * * * * * * * * *
```

```
                 Which bleach would YOU prefer?
Strongly                                                 Strongly
Prefer                         Equal                     Prefer
CURRENT:-----:-----:-----:-----:--X--:-----:-----:-----:-----: NEW
BLEACH    1     2     3     4     5     6     7     8     9  BLEACH
```

See where the X is located. Does X represent your correct answer?

PRESS NUMBER (1) FOR YES, PRESS NUMBER (2) TO GET QUESTION AGAIN.

```
         CURRENT BLEACH                    NEW BLEACH
  * * * * * * * * * * * *         * * * * * * * * * * * *
  *                      *        *                      *
  * Cost per year:       *        * Cost per year:       *
  *    $10.00            *        *      $29.00          *
  *                      *        *                      *
  * Injury level:        *        * Injury level:        *
  *  50 gas poisonings   *        *  80% decrease in     *
  *  for every           *        *  gas poisonings      *
  *  2,000,000 homes     *        *  compared to the     *
  *                      *        *  current product     *
  * * * * * * * * * * * *         * * * * * * * * * * * *
```

```
                 Which bleach would YOU prefer?
Strongly                                                 Strongly
Prefer                         Equal                     Prefer
CURRENT:-----:-----:-----:-----:--X--:-----:-----:-----:-----: NEW
BLEACH    1     2     3     4     5     6     7     8     9  BLEACH
```

See where the X is located. Does X represent your correct answer?

PRESS NUMBER (1) FOR YES, PRESS NUMBER (2) TO GET QUESTION AGAIN.

```
        CURRENT BLEACH                    NEW BLEACH
* * * * * * * * * * * *         * * * * * * * * * * * *
*                      *         *                      *
* Cost per year:       *         * Cost per year:       *
*    $10.00            *         *    $2.00             *
*                      *         *                      *
* Injury level:        *         * Injury level:        *
*   50 gas poisonings  *         *   30% INCREASE in    *
*   for every          *         *   gas poisonings     *
*   2,000,000 homes    *         *   compared to the    *
*                      *         *   current product    *
* * * * * * * * * * * *         * * * * * * * * * * * *
```

Notice that the new product has INCREASED the chance of injury!!

Which bleach would YOU prefer?

```
Strongly                                              Strongly
Prefer                          Equal                 Prefer
CURRENT:-----:-----:-----:-----:--x--:-----:-----:-----:-----: NEW
BLEACH    1     2     3     4     5     6     7     8     9  BLEACH
```

See where the X is located. Does X represent your correct answer?

PRESS NUMBER (1) FOR YES, PRESS NUMBER (2) TO GET QUESTION AGAIN.

```
        CURRENT BLEACH                    NEW BLEACH
* * * * * * * * * * * *         * * * * * * * * * * * *
*                      *         *                      *
* Cost per year:       *         * Cost per year:       *
*    $10.00            *         *    $33.00            *
*                      *         *                      *
* Injury level:        *         * Injury level:        *
*   50 gas poisonings  *         * 100% decrease in     *
*   for every          *         *   gas poisonings     *
*   2,000,000 homes    *         *   compared to the    *
*                      *         *   current product    *
* * * * * * * * * * * *         * * * * * * * * * * * *
```

Which bleach would YOU prefer?

```
Strongly                                              Strongly
Prefer                          Equal                 Prefer
CURRENT:-----:-----:-----:-----:--x--:-----:-----:-----:-----: NEW
BLEACH    1     2     3     4     5     6     7     8     9  BLEACH
```

See where the X is located. Does X represent your correct answer?

PRESS NUMBER (1) FOR YES, PRESS NUMBER (2) TO GET QUESTION AGAIN.

Thank you for answering these questions. A second type of injury can happen when bleach is used.

To find out more about it . . .
PRESS ANY KEY TO CONTINUE.

Children under 5 years old sometimes drink bleach accidently. The child then has difficulty breathing, may vomit and may complain of stomach aches. The child should be forced to vomit and then eat or drink only milk for several days.

Each year, this accident occurs to 70 out of every 2,000,000 households (the number of households in North Carolina) that use bleach. Of course, this accident is more or less likely to happen in different homes depending on how bleach is used and how often children are in the home.

Suppose you use this product in YOUR home. Would an accidental swallowing by a child be more or less likely in your home than an average home that uses bleach?

[1] Less likely   [2] More likely   [3] About the same

PRESS NUMBER (1, 2, OR 3) ON YOUR KEYBOARD.

You think an accidental swallowing of bleach is more likely in your home.

Indicate how much more likely this accident would be in your home than in the average home that uses bleach.

In YOUR home, the chance that a child swallows bleach would be:

[1]   about as likely as in an AVERAGE home
[2]   twice as likely
[4]   four times as likely
[10]   ten times as likely
[100]   one hundred times as likely

PLEASE TYPE IN A NUMBER INDICATING YOUR ANSWER. SELECT THE ONE OUT OF THE FIVE POSSIBLE CHOICES OR ANOTHER NUMBER THAT BEST ANSWERS THE QUESTION.

```
           CURRENT BLEACH                    NEW BLEACH
   * * * * * * * * * * * *        * * * * * * * * * * * *
   *                      *        *                      *
   * Cost per year:       *        * Cost per year:       *
   *    $10.00            *        *       ?              *
   *                      *        *                      *
   * Injury level:        *        * Injury level:        *
   *  70 child poisonings *        *   50% DECREASE in     *
   *  for every           *        *   child poisonings    *
   *  2,000,000 homes      *        *   compared to the     *
   *                      *        *   current product     *
   * * * * * * * * * * * *        * * * * * * * * * * * *
```

Would you be willing to pay more than $10.00 for the NEW BLEACH?

PRESS NUMBER (1) FOR YES OR NUMBER (2) FOR NO.

```
           CURRENT BLEACH                    NEW BLEACH
   * * * * * * * * * * * *        * * * * * * * * * * * *
   *                      *        *                      *
   * Cost per year:       *        * Cost per year:       *
   *    $10.00            *        *       ?              *
   *                      *        *                      *
   * Injury level:        *        * Injury level:        *
   *  70 child poisonings *        *   50% DECREASE in     *
   *  for every           *        *   child poisonings    *
   *  2,000,000 homes      *        *   compared to the     *
   *                      *        *   current product     *
   * * * * * * * * * * * *        * * * * * * * * * * * *
```

How high would the price of the NEW BLEACH have to be before you would
rather buy the CURRENT BLEACH?

$_____/year

PRESS THE NUMBERS THAT SHOW HOW MANY DOLLARS AND CENTS YOU ARE WILLING TO
PAY.

Thank you . . .

   We now want to ask you how much you prefer one bleach product over
another.

PRESS ANY KEY TO CONTINUE.

```
          CURRENT BLEACH                    NEW BLEACH
   * * * * * * * * * * * *        * * * * * * * * * * * *
   *                     *        *                     *
   * Cost per year:      *        * Cost per year:      *
   *    $10.00           *        *    $15.00           *
   *                     *        *                     *
   * Injury level:       *        * Injury level:       *
   *   70 child poisonings*        *   70% decrease in   *
   *   for every         *        *   child poisonings  *
   *   2,000,000 homes    *        *   compared to the   *
   *                     *        *   current product   *
   * * * * * * * * * * * *        * * * * * * * * * * * *
```

                  Which bleach would YOU prefer?
Strongly   Strongly
Prefer   Equal   Prefer
CURRENT:-----:-----:-----:-----:--x--:-----:-----:-----:-----: NEW
BLEACH   1   2   3   4   5   6   7   8   9   BLEACH

See where the X is located. Does X represent your correct answer?

PRESS NUMBER (1) FOR YES, PRESS NUMBER (2) TO GET QUESTION AGAIN.

```
          CURRENT BLEACH                    NEW BLEACH
   * * * * * * * * * * * *        * * * * * * * * * * * *
   *                     *        *                     *
   * Cost per year:      *        * Cost per year:      *
   *    $10.00           *        *    $19.50           *
   *                     *        *                     *
   * Injury level:       *        * Injury level:       *
   *   70 child poisonings*        *   70% decrease in   *
   *   for every         *        *   child poisonings  *
   *   2,000,000 homes    *        *   compared to the   *
   *                     *        *   current product   *
   * * * * * * * * * * * *        * * * * * * * * * * * *
```

                  Which bleach would YOU prefer?
Strongly   Strongly
Prefer   Equal   Prefer
CURRENT:-----:-----:-----:-----:--x--:-----:-----:-----:-----: NEW
BLEACH   1   2   3   4   5   6   7   8   9   BLEACH

See where the X is located. Does X represent your correct answer?

PRESS NUMBER (1) FOR YES, PRESS NUMBER (2) TO GET QUESTION AGAIN.

```
           CURRENT BLEACH                     NEW BLEACH
       * * * * * * * * * * *           * * * * * * * * * * *
       *                   *           *                   *
       * Cost per year:    *           * Cost per year:    *
       *    $10.00         *           *    $12.00         *
       *                   *           *                   *
       * Injury level:     *           * Injury level:     *
       *  70 child poisonings*         *  20% decrease in  *
       *  for every        *           *  child poisonings *
       *  2,000,000 homes  *           *  compared to the  *
       *                   *           *  current product  *
       * * * * * * * * * * *           * * * * * * * * * * *
```

                    Which bleach would YOU prefer?
Strongly   Strongly
Prefer   Equal   Prefer
CURRENT:-----:-----:-----:-----:--x--:-----:-----:-----:-----: NEW
BLEACH   1   2   3   4   5   6   7   8   9 BLEACH

See where the X is located. Does X represent your correct answer?

PRESS NUMBER (1) FOR YES, PRESS NUMBER (2) TO GET QUESTION AGAIN.


```
           CURRENT BLEACH                     NEW BLEACH
       * * * * * * * * * * *           * * * * * * * * * * *
       *                   *           *                   *
       * Cost per year:    *           * Cost per year:    *
       *    $10.00         *           *    $1.00          *
       *                   *           *                   *
       * Injury level:     *           * Injury level:     *
       *  70 child poisonings*         *  60% INCREASE in  *
       *  for every        *           *  child poisonings *
       *  2,000,000 homes  *           *  compared to the  *
       *                   *           *  current product  *
       * * * * * * * * * * *           * * * * * * * * * * *
```

      Notice that the new product has INCREASED the chance of injury!!
                    Which bleach would YOU prefer?
Strongly   Strongly
Prefer   Equal   Prefer
CURRENT:-----:-----:-----:-----:--x--:-----:-----:-----:-----: NEW
BLEACH   1   2   3   4   5   6   7   8   9 BLEACH

See where the X is located. Does X represent your correct answer?

PRESS NUMBER (1) FOR YES, PRESS NUMBER (2) TO GET QUESTION AGAIN.

```
            CURRENT BLEACH                  NEW BLEACH
     * * * * * * * * * * *        * * * * * * * * * * *
     *                   *        *                   *
     * Cost per year:    *        * Cost per year:    *
     *    $10.00         *        *    $23.00         *
     *                   *        *                   *
     * Injury level:     *        * Injury level:     *
     *  70 child poisonings*      *  100% decrease in  *
     *  for every        *        *  child poisonings  *
     *  2,000,000 homes  *        *  compared to the   *
     *                   *        *  current product   *
     * * * * * * * * * * *        * * * * * * * * * * *
```

                    Which bleach would YOU prefer?
```
Strongly                                                    Strongly
Prefer                             Equal                     Prefer
CURRENT:-----:-----:-----:-----:--X--:-----:-----:-----:-----: NEW
BLEACH    1     2     3     4     5     6     7     8     9   BLEACH
```

See where the X is located. Does X represent your correct answer?

PRESS NUMBER (1) FOR YES, PRESS NUMBER (2) TO GET QUESTION AGAIN.

     I now want to ask you some questions to help us group your responses
with the responses of others.

     How many people live in your household?

TYPE THE NUMBER OF PEOPLE IN YOUR HOUSEHOLD AND THEN PRESS THE RETURN KEY.

     How many people in your household are under 18 years old?

TYPE THE NUMBER OF PEOPLE UNDER 18 AND THEN PRESS THE RETURN KEY.

     How many people in your household are under 5 years old?

TYPE THE NUMBER OF PEOPLE UNDER 5 AND THEN PRESS THE RETURN KEY.

     How often do children under 5 visit your home?

     [1] every day
     [2] 2-3 times a week
     [3] once a week
     [4] once or twice a month
     [5] once or twice a year
     [6] never

PRESS NUMBER (1, 2, 3, 4, 5, OR 6) TO INDICATE HOW OFTEN CHILDREN VISIT.

THEN PRESS THE RETURN KEY.

What is your age?

TYPE IN YOUR AGE IN YEARS AND THEN PRESS THE RETURN KEY.

Are you now married?

PRESS (1) FOR YES OR (2) FOR NO.

What was the last grade of regular school that you completed? Do not include specialized schools like secretarial, art, or trade schools.

[1] Grade school or less (0-8)
[2] Some high school (9-11)
[3] High school graduate (12)
[4] Some college or junior college
[5] College graduate (4 or 5 year degree)
[6] Post graduate work or degree

PRESS NUMBER (1, 2, 3, 4, 5, OR 6) TO INDICATE THE LAST GRADE OF SCHOOL YOU COMPLETED.

THEN PRESS RETURN KEY.

For clarification purposes only, please tell me which category best describes the total income that you (and all other members of this household) earned DURING 1983 BEFORE TAXES. Please be sure to include each member's wages and salaries, as well as net income from any business, pensions, dividends, interest, tips, or other income.

CHOOSE THE NUMBER THAT REPRESENTS YOUR FAMILY INCOME FOR LAST YEAR.

[1] under $10,000
[2] $10,000 to less than $20,000
[3] $20,000 to less than $30,000
[4] $30,000 to less than $40,000
[5] $40,000 to less than $50,000
[6] $50,000 to less than $60,000
[7] $60,000 to less than $70,000
[8] $70,000 to less than $80,000
[9] $80,000 and over

This information is completely confidential and cannot be traced to you in any way.

However, if you feel uncomfortable answering this question, just type the number (0).

NOW, PRESS THE NUMBER THAT BEST REPRESENTS YOUR FAMILY'S INCOME.
THEN PRESS THE RETURN KEY.

Thank you for helping with our project.

Please tell the computer monitor you are finished.

Thanks again! I enjoyed working with you! Bye!

APPENDIX B

Drain Opener Questionnaire

SUBJECT # _____

INTERVIEWER NAME _____

Introduce yourself and ask subject if he/she is ready to begin. Go to Stage 1.

Stage 1: Examination of Drain Opener

(Show subject the liquid drain opener.)

Please examine this new drain opener as if you are about to use it in your home for the first time. I will ask you questions about the way you would use it. There are no right or wrong answers to any of the questions, just answer them as honestly as you can.

Take as much time as you would if you were going to use it in your home.

Stage 2: Questions about Drain Opener Usage

While using this product, is it likely that you would:

2.1 Pour through standing water to unclog drain	yes	no	n/a
2.2 Pour through pipes connected to a septic tank	yes	no	n/a
2.3 Use with a plunger to help unclog the drain	yes	no	n/a
2.4 Wear rubber gloves	yes	no	n/a

Stage 3: Storage of the Product

3.1 Where would you store a product like this in your home?

(Do not read. Probe if necessary.
    In a childproof location? _____ Yes   _____ No)

(Acceptable probes:
    Is that above or below your washing machine?
    Is that above or below the counter?)

Stage 4: Precautions Valuation--Drain Opener

Drain opener is made a number of different ways. Presently, most products on the market have some potential drawbacks as well as benefits. For example, many drain openers have a bitter smell. To help us know which are most valuable to you, I am going to name a number of possible improvements in a $1.79 bottle of drain opener. For each tell me how much more, if anything, you would be willing to pay for the new drain opener.

Current product	New product	What is the most you would pay OVER $1.79 for the new product?
4.1 Bitter smell	Pleasant lemony scent	_____ cents
4.2 Container is difficult to open	Easy-open container	_____ cents
4.3 Cannot be used in toilets	Can be used in toilets	_____ cents
4.4 Must be stored on a high shelf	Can be stored anywhere	_____ cents
4.5 Gloves must be worn when using	Gloves need not be worn when using	_____ cents

Stage 5: Frequency of Use Questions

5.1 Do you currently have any liquid drain opener in your house?

(Yes _____)   (No _____)

(If no, then ask:
    Do you use either dry, crystallized drain openers or lye?)
        (Yes _____)   (No _____)

5.2 How many bottles of liquid drain opener equivalent to the one we showed
    you does your household use in a given year?

    (ask by month if necessary and then convert)
    (bottles of drain opener/year _____?)

Stage 6: Introduction to the Personal Computer

Now let's turn to the personal computer and let you respond to
questions from it. Each time a question appears on the screen, please
read it and then answer the question. The computer will ask you to press
a number on the keyboard that corresponds to your answer. (Show subject
numeric keyboard.) Then you press the return key so the computer knows
that you have answered the questions.
(Show subject the return key.)

(If first question does not show up on the screen, insert disk and
turn on the computer. The screen will show a question asking the subject
to input his subject number. Tell subject his subject number and let him
type it into the computer.)

(Stay with the subject until he correctly answers the first question
asking "What month is it?".)

I will leave you to answer the remaining questions yourself. If you
have questions, there is someone here to help you. (Point out the
computer reference person.) Just raise you hand and she will be right
over. When you finish, tell me and I will answer any questions you might
have. (Thank subject and leave him at personal computer.)

BEFORE WE BEGIN, COULD YOU ENTER YOUR SUBJECT NUMBER (THIS IS A NUMBER
GIVEN TO YOU BY THE PROGRAM OPERATOR).

AFTER TYPING IN YOUR NUMBER, PLEASE HIT THE RETURN KEY.

THANK YOU!

Hello! This is your first question on the computer.

What month is it?

[1] May   [2] June   [3] July

PRESS A NUMBER (1, 2, OR 3) INDICATING THE MONTH.

THEN PRESS THE RETURN KEY.

Well done! You are correct. It is June.

PRESS ANY KEY TO CONTINUE.

If liquid drain opener splashes onto someone's hands, it causes
painful burns or red, swollen blisters. The treatment is to see a
doctor, who will carefully wash the injured hands. Complete healing
occurs within a week.

PRESS ANY KEY TO CONTINUE.

This accident occurs to 45 out of 2,000,000 households (the number of households in North Carolina) which use drain opener. Of course, this accident is more or less likely to happen in different homes depending on how liquid drain opener is used.

Suppose you used this product in YOUR home. Would the chance of an accidental hand burn be more or less likely in your home than in an average home that uses liquid drain opener?

[1] Less likely   [2] More likely   [3] About the same

PRESS NUMBER (1, 2, OR 3) ON YOUR KEYBOARD.

This means that you think an accidental hand burn is more likely in an average home than in your home.

Indicate how much more likely this accident is in an average home that uses liquid drain opener than in your home.

In an AVERAGE home, a hand burn from splashing liquid drain opener would be:

    [1]   about as likely as in YOUR home
    [2]   twice as likely
    [4]   four times as likely
    [10]   ten times as likely
    [100]   one hundred times as likely

PLEASE TYPE IN A NUMBER INDICATING YOUR ANSWER. SELECT ONE OF THE FIVE POSSIBLE CHOICES OR ANOTHER NUMBER THAT BEST ANSWERS THE QUESTION.

The following questions will ask you to compare two liquid drain opener products. The CURRENT DRAIN OPENER, now sold in stores, is always listed on the left side of your screen. The NEW DRAIN OPENER is always found on the right side of your screen.

Let's try a practice question first.
PRESS ANY KEY TO CONTINUE.

```
       CURRENT DRAIN OPENER              NEW DRAIN OPENER
     * * * * * * * * * * *          * * * * * * * * * * * *
     *                   *          *                    *
     * Cost per year:    *          * Cost per year:     *
     *    $7.00          *          *        ?           *
     *                   *          *                    *
     * Injury level:     *          * Injury level:      *
     *  45 hand burnings *          *  50% DECREASE in    *
     *  for every        *          *  hand burnings     *
     *  2,000,000 homes  *          *  compared to the   *
     *                   *          *  current product   *
     * * * * * * * * * * *          * * * * * * * * * * * *
```

    Let's make certain that you understand the question completely. The
two liquid drain opener products are IDENTICAL IN ALL RESPECTS except for
their chances of causing hand burns. The NEW DRAIN OPENER is 50% less
likely to result in hand burns in your home than the CURRENT DRAIN
OPENER.

    The amount you are willing to pay will affect only YOUR household's
costs and not those of any other liquid drain opener users.

READY TO ANSWER THE QUESTION? PRESS ANY KEY TO CONTINUE.

```
       CURRENT DRAIN OPENER              NEW DRAIN OPENER
     * * * * * * * * * * *          * * * * * * * * * * * *
     *                   *          *                    *
     * Cost per year:    *          * Cost per year:     *
     *    $ 7.00         *          *        ?           *
     *                   *          *                    *
     * Injury level:     *          * Injury level:      *
     *  45 hand burnings *          *  50% DECREASE in    *
     *  for every        *          *  hand burnings     *
     *  2,000,000 homes  *          *  compared to the   *
     *                   *          *  current product   *
     * * * * * * * * * * *          * * * * * * * * * * * *
```

    Would you be willing to pay more than $7.00 for the NEW DRAIN
OPENER?

PRESS NUMBER (1) FOR YES OR NUMBER (2) FOR NO.

```
        CURRENT DRAIN OPENER            NEW DRAIN OPENER
   * * * * * * * * * * *        * * * * * * * * * * * *
   *                   *        *                      *
   * Cost per year:    *        * Cost per year:       *
   *    $ 7.00         *        *        ?             *
   *                   *        *                      *
   * Injury level:     *        * Injury level:        *
   *  45 hand burnings *        *  50% DECREASE in      *
   *  for every        *        *  hand burnings        *
   *  2,000,000 homes  *        *  compared to the      *
   *                   *        *  current product      *
   * * * * * * * * * * *        * * * * * * * * * * * *
```

　　　　How high would the price of the NEW DRAIN OPENER have to be before
you would rather buy the CURRENT DRAIN OPENER?

$ _____/year

PRESS THE NUMBERS THAT SHOW HOW MANY DOLLARS AND CENTS YOU ARE WILLING TO
PAY.

Thank you . . .

　　　　We now want to ask you how much you prefer one liquid drain opener
product over another.

PRESS ANY KEY TO CONTINUE.

```
        CURRENT DRAIN OPENER              NEW DRAIN OPENER
    * * * * * * * * * * *             * * * * * * * * * * * *
    *                   *             *                     *
    * Cost per year:    *             * Cost per year:      *
    *     $ 7.00        *             *       $11.00        *
    *                   *             *                     *
    * Injury level:     *             * Injury level:       *
    *   45 hand burnings*             *   40% decrease in    *
    *   for every       *             *   hand burnings     *
    *   2,000,000 homes *             *   compared to the   *
    *                   *             *   current product   *
    * * * * * * * * * * *             * * * * * * * * * * * *
```

                Which liquid drain opener would YOU prefer?

```
Strongly                                                   Strongly
Prefer                           Equal                     Prefer
CURRENT:-----:-----:-----:-----:--X--:-----:-----:-----:-----: NEW
DRAIN     1     2     3     4     5     6     7     8     9  DRAIN
OPENER                                                      OPENER
```

See where the X is located. Does X represent your correct answer?

PRESS NUMBER (1) FOR YES, PRESS NUMBER (2) TO GET QUESTION AGAIN.

```
        CURRENT DRAIN OPENER              NEW DRAIN OPENER
    * * * * * * * * * * *             * * * * * * * * * * * *
    *                   *             *                     *
    * Cost per year:    *             * Cost per year:      *
    *     $ 7.00        *             *       $14.00        *
    *                   *             *                     *
    * Injury level:     *             * Injury level:       *
    *   45 hand burnings*             *   40% decrease in    *
    *   for every       *             *   hand burnings     *
    *   2,000,000 homes *             *   compared to the   *
    *                   *             *   current product   *
    * * * * * * * * * * *             * * * * * * * * * * * *
```

                Which liquid drain opener would YOU prefer?
```
Strongly                                                   Strongly
Prefer                           Equal                     Prefer
CURRENT:-----:-----:-----:-----:--X--:-----:-----:-----:-----: NEW
DRAIN     1     2     3     4     5     6     7     8     9  DRAIN
OPENER                                                      OPENER
```

See where the X is located. Does X represent your correct answer?

PRESS NUMBER (1) FOR YES, PRESS NUMBER (2) TO GET QUESTION AGAIN.

```
        CURRENT DRAIN OPENER              NEW DRAIN OPENER
      * * * * * * * * * * *           * * * * * * * * * * * *
      *                   *           *                      *
      * Cost per year:    *           * Cost per year:       *
      *    $ 7.00         *           *       $21.00         *
      *                   *           *                      *
      * Injury level:     *           * Injury level:        *
      *  45 hand burnings *           *  80% decrease in      *
      *  for every        *           *  hand burnings        *
      *  2,000,000 homes  *           *  compared to the      *
      *                   *           *  current product      *
      * * * * * * * * * * *           * * * * * * * * * * * *
```

           Which liquid drain opener would YOU prefer?
Strongly   Strongly
Prefer   Equal   Prefer
CURRENT:----:-----:-----:-----:--x--:-----:-----:-----:-----: NEW
DRAIN   1   2   3   4   5   6   7   8   9 DRAIN
OPENER   OPENER

See where the X is located. Does X represent your correct answer?

PRESS NUMBER (1) FOR YES, PRESS NUMBER (2) TO GET QUESTION AGAIN.

```
        CURRENT DRAIN OPENER              NEW DRAIN OPENER
      * * * * * * * * * * *           * * * * * * * * * * * *
      *                   *           *                      *
      * Cost per year:    *           * Cost per year:       *
      *    $ 7.00         *           *       $ 1.00         *
      *                   *           *                      *
      * Injury level:     *           * Injury level:        *
      *  45 hand burnings *           *  30% INCREASE in      *
      *  for every        *           *  hand burnings        *
      *  2,000,000 homes  *           *  compared to the      *
      *                   *           *  current product      *
      * * * * * * * * * * *           * * * * * * * * * * * *
```

     Notice that the new product has INCREASED the chance of injury!!
           Which liquid drain opener would YOU prefer?
Strongly   Strongly
Prefer   Equal   Prefer
CURRENT:----:-----:-----:-----:--x--:-----:-----:-----:-----: NEW
DRAIN   1   2   3   4   5   6   7   8   9 DRAIN
OPENER   OPENER

See where the X is located. Does X represent your correct answer?

PRESS NUMBER (1) FOR YES, PRESS NUMBER (2) TO GET QUESTION AGAIN.

```
        CURRENT DRAIN OPENER            NEW DRAIN OPENER
      * * * * * * * * * * *         * * * * * * * * * * * *
      *                   *         *                      *
      * Cost per year:    *         * Cost per year:       *
      *    $ 7.00         *         *      $24.00          *
      *                   *         *                      *
      * Injury level:     *         * Injury level:        *
      *   45 hand burnings *         * 100% decrease in     *
      *   for every       *         *   hand burnings      *
      *   2,000,000 homes  *         *   compared to the    *
      *                   *         *   current product    *
      * * * * * * * * * * *         * * * * * * * * * * * *
```

           Which liquid drain opener would YOU prefer?

```
Strongly                                              Strongly
Prefer                         Equal                  Prefer
CURRENT:-----:-----:-----:-----:--x--:-----:-----:-----:-----: NEW
DRAIN     1     2     3     4     5     6     7     8     9   DRAIN
OPENER                                                       OPENER
```

See where the X is located. Does X represent your correct answer?

PRESS NUMBER (1) FOR YES, PRESS NUMBER (2) TO GET QUESTION AGAIN.

     Thank you for answering these questions. A second type of injury
can happen when liquid drain opener is used.

To find out more about it . . .
PRESS ANY KEY TO CONTINUE.

     Children under 5 years old sometimes drink liquid drain opener. The
result can be severe painful burns to the mouth and throat. An operation
may be needed to replace parts of the throat, and hospital treatment may
last up to 3 weeks.

     Each year, this accident occurs to 35 out of every 2,000,000
households (the number of households in North Carolina) that use liquid
drain opener. Of course, this accident is more or less likely to happen
in different homes depending on how liquid drain opener is used and how
often children are in the home.

Suppose you use this product in YOUR home. Would an accidental swallowing by a child be more or less likely in your home than in an average home that uses liquid drain opener?

[1] Less likely   [2] More likely   [3] About the same

PRESS NUMBER (1, 2, OR 3) ON YOUR KEYBOARD.

You think an accidental swallowing of liquid drain opener is more likely in your home.

Indicate how much more likely this accident would be in your home than in the average home that uses liquid drain opener.

In YOUR home, the chance that a child swallows liquid drain opener would be:

[1]   about as likely as in an AVERAGE home
[2]   twice as likely
[4]   four times as likely
[10]   ten times as likely
[100]   one hundred times as likely

PLEASE TYPE IN A NUMBER INDICATING YOUR ANSWER. SELECT THE ONE OUT OF THE FIVE POSSIBLE CHOICES OR ANOTHER NUMBER THAT BEST ANSWERS THE QUESTION.

```
        CURRENT DRAIN OPENER           NEW DRAIN OPENER
    * * * * * * * * * * *          * * * * * * * * * * * *
    *                   *          *                      *
    * Cost per year:    *          * Cost per year:       *
    *    $ 7.00         *          *       ?              *
    *                   *          *                      *
    * Injury level:     *          * Injury level:        *
    *  35 child poisonings*        *  50% DECREASE in      *
    *  for every        *          *  child poisonings     *
    *  2,000,000 homes  *          *  compared to the      *
    *                   *          *  current product      *
    * * * * * * * * * * *          * * * * * * * * * * * *
```

Think about how much you would be willing to pay for the liquid drain opener on the right.

Would you be willing to pay more than $7.00 for the NEW DRAIN OPENER?

PRESS NUMBER (1) FOR YES OR NUMBER (2) FOR NO.

```
      CURRENT DRAIN OPENER              NEW DRAIN OPENER
    * * * * * * * * * * *           * * * * * * * * * * * *
    *                   *           *                      *
    * Cost per year:    *           * Cost per year:       *
    *    $ 7.00         *           *        ?             *
    *                   *           *                      *
    * Injury level:     *           * Injury level:        *
    *  35 child poisonings*         *   50% DECREASE in     *
    *  for every        *           *   child poisonings    *
    *  2,000,000 homes  *           *   compared to the     *
    *                   *           *   current product     *
    * * * * * * * * * * *           * * * * * * * * * * * *
```

How high would the price of the NEW DRAIN OPENER have to be before you
would rather buy the CURRENT DRAIN OPENER?

$_____/year

PRESS THE NUMBERS THAT SHOW HOW MANY DOLLARS AND CENTS YOU ARE WILLING TO
PAY.

Thank you . . .
    We now want to ask you how much you prefer one liquid drain opener
product over another.

PRESS ANY KEY TO CONTINUE.


```
      CURRENT DRAIN OPENER              NEW DRAIN OPENER
    * * * * * * * * * * *           * * * * * * * * * * * *
    *                   *           *                      *
    * Cost per year:    *           * Cost per year:       *
    *    $ 7.00         *           *     $11.00           *
    *                   *           *                      *
    * Injury level:     *           * Injury level:        *
    *  35 child poisonings*         *   60% decrease in     *
    *  for every        *           *   child poisonings    *
    *  2,000,000 homes  *           *   compared to the     *
    *                   *           *   current product     *
    * * * * * * * * * * *           * * * * * * * * * * * *
```

Which liquid drain opener would YOU prefer?

```
Strongly                                              Strongly
Prefer                          Equal                 Prefer
CURRENT:-----:-----:-----:-----:--X--:-----:-----:-----:-----: NEW
DRAIN      1     2     3     4     5     6     7     8     9  DRAIN
OPENER                                                      OPENER
```

See where the X is located. Does X represent your correct answer?

PRESS NUMBER (1) FOR YES, PRESS NUMBER (2) TO GET QUESTION AGAIN.

```
        CURRENT DRAIN OPENER          NEW DRAIN OPENER
    * * * * * * * * * * *        * * * * * * * * * * * *
    *                   *        *                      *
    * Cost per year:    *        * Cost per year:       *
    *    $ 7.00         *        *       $ 9.00         *
    *                   *        *                      *
    * Injury level:     *        * Injury level:        *
    * 35 child poisonings*       *   20% decrease in    *
    * for every         *        *   child poisonings   *
    * 2,000,000 homes    *       *   compared to the    *
    *                   *        *   current product    *
    * * * * * * * * * * *        * * * * * * * * * * * *
```

            Which liquid drain opener would YOU prefer?

```
Strongly                                                Strongly
Prefer                         Equal                    Prefer
CURRENT:-----:-----:-----:-----:--x--:-----:-----:-----:-----: NEW
DRAIN      1     2     3     4     5     6     7     8     9  DRAIN
OPENER                                                      OPENER
```

See where the X is located. Does X represent your correct answer?

PRESS NUMBER (1) FOR YES, PRESS NUMBER (2) TO GET QUESTION AGAIN.

```
        CURRENT DRAIN OPENER          NEW DRAIN OPENER
    * * * * * * * * * * *        * * * * * * * * * * * *
    *                   *        *                      *
    * Cost per year:    *        * Cost per year:       *
    *    $ 7.00         *        *       $ 0.50         *
    *                   *        *                      *
    * Injury level:     *        * Injury level:        *
    * 35 child poisonings*       *   60% INCREASE in    *
    * for every         *        *   child poisonings   *
    * 2,000,000 homes    *       *   compared to the    *
    *                   *        *   current product    *
    * * * * * * * * * * *        * * * * * * * * * * * *
```

    Notice that the new product has INCREASED the chance of injury!!

            Which liquid drain opener would YOU prefer?

```
Strongly                                                Strongly
Prefer                         Equal                    Prefer
CURRENT:-----:-----:-----:-----:--x--:-----:-----:-----:-----: NEW
DRAIN      1     2     3     4     5     6     7     8     9  DRAIN
OPENER                                                      OPENER
```

See where the X is located. Does X represent your correct answer?

PRESS NUMBER (1) FOR YES, PRESS NUMBER (2) TO GET QUESTION AGAIN.

```
        CURRENT DRAIN OPENER              NEW DRAIN OPENER
      * * * * * * * * * * * *          * * * * * * * * * * * *
      *                      *         *                      *
      * Cost per year:       *         * Cost per year:       *
      *      $ 7.00          *         *        $21.00        *
      *                      *         *                      *
      * Injury level:        *         * Injury level:        *
      *  35 child poisonings *         * 100% decrease in      *
      *  for every           *         *  child poisonings    *
      *  2,000,000 homes      *        *  compared to the     *
      *                      *         *  current product     *
      * * * * * * * * * * * *          * * * * * * * * * * * *

         Which liquid drain opener would YOU prefer?
```

```
Strongly                                                Strongly
Prefer                          Equal                   Prefer
CURRENT:-----:-----:-----:-----:--X--:-----:-----:-----: NEW
DRAIN     1     2     3     4     5     6     7     8     9   DRAIN
OPENER                                                       OPENER
```

See where the X is located. Does X represent your correct answer?

PRESS NUMBER (1) FOR YES, PRESS NUMBER (2) TO GET QUESTION AGAIN.

     I now want to ask you some questions to help us group your responses
with the responses of others.

     How many people live in your household?

TYPE THE NUMBER OF PEOPLE IN YOUR HOUSEHOLD AND THEN PRESS THE RETURN
KEY.

     How many people in your household are under 18 years old?

TYPE THE NUMBER OF PEOPLE UNDER 18 AND THEN PRESS THE RETURN KEY.

How many people in your household are under 5 years old?

TYPE THE NUMBER OF PEOPLE UNDER 5 AND THEN PRESS THE RETURN KEY.

How often do children under 5 visit your home?

[1] every day
[2] 2-3 times a week
[3] once a week
[4] once or twice a month
[5] once or twice a year
[6] never

PRESS NUMBER (1, 2, 3, 4, 5, OR 6) TO INDICATE HOW OFTEN CHILDREN VISIT.

THEN PRESS THE RETURN KEY.

What is your age?

TYPE IN YOUR AGE IN YEARS AND THEN PRESS THE RETURN KEY.

Are you now married?

PRESS (1) FOR YES OR (2) FOR NO.

What was the last grade of regular school that you completed? Do not include specialized schools like secretarial, art, or trade schools.

[1] Grade school or less (0-8)
[2] Some high school (9-11)
[3] High school graduate (12)
[4] Some college or junior college
[5] College graduate (4 or 5 year degree)
[6] Post graduate work or degree

PRESS NUMBER (1, 2, 3, 4, 5, OR 6) TO INDICATE THE LAST GRADE OF SCHOOL YOU COMPLETED.

THEN PRESS RETURN KEY.

For clarification purposes only, please tell me which category best describes the total income that you (and all other members of this household) earned DURING 1983 BEFORE TAXES. Please be sure to include each member's wages and salaries, as well as net income from any business, pensions, dividends, interest, tips, or other income.

CHOOSE THE NUMBER THAT REPRESENTS YOUR FAMILY INCOME FOR LAST YEAR.

    [1] under $10,000
    [2] $10,000 to less than $20,000
    [3] $20,000 to less than $30,000
    [4] $30,000 to less than $40,000
    [5] $40,000 to less than $50,000
    [6] $50,000 to less than $60,000
    [7] $60,000 to less than $70,000
    [8] $70,000 to less than $80,000
    [9] $80,000 and over

This information is completely confidential and cannot be traced to you in any way.

However, if you feel uncomfortable answering this question, just type the number (0).

NOW, PRESS THE NUMBER THAT BEST REPRESENTS YOUR FAMILY'S INCOME.
THEN PRESS THE RETURN KEY.

Thank you for helping with our project.

Please tell the computer monitor you are finished.

Thanks again! I enjoyed working with you! Bye!

APPENDIX C

Variation in Consumers' Valuation of Product Attributes

As was noted in Section 4.4, consumers' mean valuations of the precautionary actions for bleach and drain opener ranged from 15 cents to 19 cents per container. Although these values may accurately reflect consumer preferences, an alternative hypothesis is that consumers did not make refined judgments about the value of avoiding taking these precautions. Instead, they may have simply indicated that any beneficial product attribute must be worth about 15 or 20 cents.

There are two bases for rejecting the uniform response hypothesis. First, responses were not identical across attribute categories. The mean willingness to pay for drain opener attributes ranged from 9 cents to 30 cents per container, and the standard deviations of these responses were substantial (see Table C.1). The cleaning agent attributes also received quite different valuations, ranging from 14 to 26 cents per container. If consumers gave random answers or identical answers for all attributes, the mean response levels would have been more similar.

A second indication that the responses were accurate reflections of tastes would be a very strong correlation among the contingent valuation responses not only for the two risk-related attributes but also for the other product attributes included in the survey. Table C.2 presents the correlation matrix for the bleach contingent valuation responses, Table C.3 the correlation matrix for the drain opener contingent valuation responses. In the case of bleach, the five product attributes for which contingent valuation responses were obtained were piney scent, reusable container, cap measure, storage anywhere, and combinability with other cleaners. The correlation coefficients range from 0.26 to 0.84. For the drain opener contingent valuation responses, the product attributes were lemony scent, easy-open container, use in toilets, storage anywhere, and no need to wear gloves. In this case, too, there is a wide range of correlation coefficients, from 0.06 to 0.82.

The absolute levels of these correlations are not great enough to allow us to conclude that there was a similar uniform response to

Table C.1 Means and standard deviations for willingness to pay for product attributes

Product attribute	Mean dollars per container (std. deviation)
Drain opener	
Lemony scent	0.19
	(0.32)
Easy-open container	0.09
	(0.23)
Use in toilets	0.30
	(0.54)
Storage anywhere	0.15
	(0.33)
No need to wear gloves	0.17
	(0.34)
Bleach	
Piney scent	0.26
	(0.52)
Reusable container	0.18
	(0.53)
Measuring cap	0.14
	(0.26)
Storage anywhere	0.16
	(0.44)
Combinability with other cleaners	0.19
	(0.46)

all the questions. Moreover, the substantial variation in the correlation among these different attributes' valuations indicates that within some pairs of product attributes there is a much stronger relationship than within other pairs. Each of these findings is consistent with the fundamental assumption underlying the analysis, which is that consumers did give reasoned responses to the contingent valuation questions and did not simply offer to pay some uniform amount for any positively valued attribute regardless of its identity.

Table C.2 Bleach contingent valuation correlation matrix

Product attribute	Piney scent	Reusable container	Cap measure	Storage anywhere	Combinability with other cleaners
Piney scent	1.00	0.84	0.33	0.31	0.30
Reusable container		1.00	0.26	0.31	0.32
Cap measure			1.00	0.64	0.70
Storage anywhere				1.00	0.60
Combinability with other cleaners					1.00

Table C.3 Drain opener contingent valuation correlation matrix

Product attribute	Lemony scent	Easy-open container	Use in toilets	Storage anywhere	No need to wear gloves
Lemony scent	1.00	0.26	0.08	0.06	0.17
Easy-open container		1.00	0.30	0.32	0.32
Use in toilets			1.00	0.33	0.28
Storage anywhere				1.00	0.82
No need to wear gloves					1.00

APPENDIX D

Effect of Labels on Other Consumer Actions

Ideally, hazard warnings should produce a consumer response targeted to the particular precautions indicated. Warnings of a more generally alarmist nature may also lead consumers to alter actions unrelated to the warnings on the label. Such distortions are undesirable from the standpoint of the consumer, since there are costs associated with altering actions that would have been desirable in the absence of the label.

In the case of bleach, the other product uses pertained to whether or not the bleach could be used to clean sinks, to clean floors, to remove mildew, and to remove problem stains. With the possible exception of cleaning floors, these uses of bleach are very common and ideally the label should not discourage consumers from undertaking such activities.

Table D.1 summarizes the coefficients for the effects of the different labels on these four actions. In each case the coefficients were estimated as part of a more elaborate precaution probability equation, such as those in Table 4.4. Both the Bright and the Test label are less likely to lead consumers to think that they can use the cleaning agent to clean sinks or floors. These differential responses to the label are a consequence of the different information provided on product uses. Cleaning sinks and floors was not a specifically indicated use of the product for Bright and Test, whereas both the Clorox and the label with no warning featured these uses quite prominently. This pattern of responses is exactly what would be expected if consumers processed the labeling information reliably.

Two fundamental purposes of any bleach product are to remove mildew and problem stains. These uses were featured prominently on all the labels, and there was no differential effect among the labels on these precautions.

Table D.2 presents similar results for the WARNAREA estimates, analyzing the effect of the size of the hazard warning rather than relying on 0–1 dummy variables. Increasing the area of the label decreased the use of the product for cleaning sinks and floors and

Table D.1 Bleach label effects on actions other than precautions[a]

	Coefficients (standard errors)			
Label	Clean dirty sinks	Clean floors	Remove mildew	Remove problem stains
Clorox	0.189	−0.264	−0.032	0.409
	(0.460)	(0.468)	(0.434)	(0.403)
Bright	−0.783	−1.284	−0.251	−0.307
	(0.463)	(0.480)	(0.469)	(0.430)
Test	−0.799	−1.244	−0.141	−0.021
	(0.499)	(0.473)	(0.458)	(0.419)

a. Results are based on logit equations including all the variables used in Table 4.4 except *FIVE*.

had no effect on the propensity to use the product either to remove mildew or to wash problem stains.

The drain opener labels did not lead consumers to alter actions other than those specified in the precautions on the label. Each of the three actions specified—pouring the drain opener through standing water to unclog a drain, pouring the drain opener through pipes connected to a septic tank, and using the drain opener with

Table D.2 WARNAREA effects on bleach uses[a]

Use	WARNAREA coefficients (asymptotic std. errors)
Clean sinks	−0.0134
	(0.0066)
Clean floors	−0.0197
	(0.0068)
Remove mildew	−0.0024
	(0.0065)
Remove problem stains	−0.0024
	(0.0065)
Remove problem stains	−0.0016
	(0.0060)

a. Coefficients are estimated on the basis of equations that parallel those in Table 4.7.

Table D.3 Drain opener label effects on actions other than precautions[a]

	Coefficients (asymptotic std. errors)		
Label	Unclog drain through standing water	Unclog pipe to septic tank	Unclog drain with plunger
Drāno	−0.449	−0.189	−0.640
	(0.391)	(0.458)	(0.404)
Test	0.151	0.266	−0.400
	(0.407)	(0.449)	(0.413)

a. Results are based on logit equations including the same set of variables used in Table 4.5.

a plunger to help unclog the drain—posed no physical risk but did relate to proper use of the product.

Table D.3 shows the two labeling variables had no significant effects on these actions. All the labels advised against pouring the product through standing water, and there appears to be no differential effectiveness. Similarly, all products can be used if there is a septic tank, and there are no significant effects of label format on this action, which is the result one would hope to find. Both the Drāno and Test label advised against using the drain opener with a plunger, and in each case the expected negative influence is observed. Only in the Drāno case does the coefficient exceed its standard error ($t = 1.58$). While not statistically significant at the 5 percent level, it is noteworthy that the stongest effects of the Drāno and Test labels occur in the only case in which a differential impact of the label was expected.

Table D.4 shows the results of a similar test using the WARN-AREA variable rather than the label dummy variables. Whereas WARNAREA has a positive and significant effect on the two main variables of interest—wearing rubber gloves and storage in a child-proof location—none of the effects for other uses is statistically significant. The results for plunger use are the closest to passing such a test, as in the label dummy variable results.

For both products, there is surprisingly little evidence of a spill-over effect of labels on other uses the product might have. The risk information provided stimulates a highly focused response and does not lead to any apparent general distortion in the product uses that consumers envision for either bleach or drain opener.

Table D.4 WARNAREA effects on drain opener uses[a]

Use	WARNAREA coefficients (asymptotic std. errors)
Unclog drain through standing water	−0.0037
	(0.0048)
Unclog pipe to septic tank	−0.0005
	(0.0055)
Unclog drain with a plunger	−0.0077
	(0.0049)

a. Results are based on logit equations that parallel those in Table 4.7.

APPENDIX E
Linearity of Rating Scale

To test the linearity of the response scale, a necessary condition in using ordinary least-squares regression for estimating the utility weights on costs and injury reductions, we used Forrest Young's ALSCAL algorithm (1981) to find the monotone transformation of the rating scale that maximizes its fit with the linear specification of the attribute weighting function (that is, the right side of equation 5.1). Figure E.1 graphs the optimally transformed scale against the original scale. The transformed scale, though not perfectly linear, does not appear to fit any clear pattern with respect to the deviations from linearity. Furthermore, the coefficients from the optimal run correlate at 0.96 with the uncorrected values. Thus, we simply report the results from the linear formulation in Table 5.1.

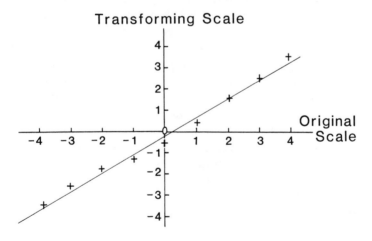

Figure E.1 Optimal transformation of rating scale using alternating least squares.

APPENDIX F

Additional Results about Injury Valuations

Because the paired-comparison questions consisted of six questions asked of each subject, they allowed us to test whether responses to variations in the differences in risk between the paired products depended upon two specific characteristics of these comparisons. First, we hypothesized that subjects would pay a certainty premium: that is, they would pay more for a product that totally eliminates the hazards from use than for a new product that reduces but does not totally eliminate the hazard. Second, we hypothesized that households are willing to pay more to eliminate an increase in the injury rate than to reduce the injury rate. This last conjecture is consistent with the Kahneman and Tversky (1979) prospect theory, in which the utility function for avoiding losses is steeper than the utility function for gains.

To test the certainty premium hypothesis, we again used conjoint analysis to analyze all the paired comparison questions, and we used regression analysis to analyze the subset of indifferent paired comparisons. The conjoint approach showed that 62 percent of the subjects attached a higher value per (hand burn and chloramine gassing) injury reduced when the new product totally eliminated the risks from the new product than when the product only partially eliminated the risks. The remaining 38 percent attached a lower value per injury reduced when the new product totally eliminated all injuries. To obtain a sense of the order of magnitude of the distribution of these certainty premiums across all subjects, we calculated the mean increase in the value of an injury to be 9 percent for products that totally eliminate the hazards from their use.

Column 1 of Table F.1 displays the certainty premiums (above the injury valuations in column 4 of Table 5.1) derived from the indifferent product pairs. The estimates come from a linear regression of the difference in cost on the difference in risk, plus a cross-product term that multiplies the risk difference by a dummy variable taking the value of 1 when the new product is risk free. All four coefficients on the cross-product term indicate a positive risk

Table F.1 Morbidity valuations from comparison of equally valued pairs of products[a]

| | (1) | (2) | (3) | (4) |
| | $/injury due to certainty | | $/injury due to reversal | |
	Linear specification	Quadratic specification	Linear specification	Quadratic specification
Bleach				
Chloramine gas	0.26	0.02	−0.36	0.30
	(1.12)	(0.06)	(−0.46)	(0.30)
Child poisoning	0.25	0.32	−0.23	−0.57
	(2.21)	(1.34)	(−1.14)	(−0.68)
Drain opener				
Hand burns	0.15	0.13	−0.20	−0.27
	(2.21)	(0.74)	(−0.52)	(0.61)
Child poisoning	0.32	0.00	−0.47	0.65
	(2.44)	(0.99)	(−1.55)	(0.65)

Note: Each subject's morbidity valuation was calculated by dividing the difference in cost between the old and new products by their difference in the number of injuries.

a. All values are measured in dollars per injury. *t*-statistics are given in parentheses.

premium with magnitudes between 25 and 90 percent of the injury values in column 4 of Table 5.1. In order to test the sensitivity of these certainty premium estimates to the linear specification of the relationship between differences in cost and risk, we also estimated a quadratic relationship. Column 2 presents the estimates of the certainty premiums derived from adding a cross-product term to the quadratic specifiction. Three out of the four certainty premiums in column 2 are less than those in column 1, although they all show positive certainty premiums. This result suggests that while the magnitude of the true certainty premiums may be less than those given by the linear specification (column 1), they are still likely to be positive.

The analyses of the paired comparison data show much weaker support for the second hypothesis, that households value avoiding risk increases more than reducing risk. From the conjoint analysis of all the paired comparisons, 58 percent of subjects were willing to pay more for the elimination of an increase in the (hand burn or gassing) injury rate by one injury than for the reduction in the injury rate by one injury. Again, to give some idea of the magnitudes of the distribution of these effects, we estimated the mean amount by which the value of avoiding higher injury rates exceeded the value of reducing injury rates to be 14 percent. In contrast, the indifferent pairs analyses provide no support for the risk reversal hypotheses. Again we proceeded by adding a cross-product term to the regression equation of risk differences on cost differences, with the cross-product term formed by multiplying the risk difference by a dummy variable that equals 1 when the risk of the new product exceeds the risk of the current product. The linear specification suggests that the injury valuations are *less* for risk increases than for decreases, contrary to our hypothesis (see column 3). However, the quadratic specification shows no systematic effect of the risk reversal (see column 4), and thus provides a possible explanation for the discrepancy between the linear specification results and the conjoint results.

The findings concerning these last two hypotheses should caution policymakers that consumers' valuations of morbidity benefits are sensitive to the form in which these preferences are elicited. Specifically, consumers' morbidity valuations appear to depend upon whether or not the product attribute valued results in a total elimination of injuries and may depend upon whether the choice situation involves valuing the avoidance of an increase in the injury rate or valuing the decrease in the injury rate.

Having compared the mean CV values for injury valuations with the mean conjoint values and to the indifferent pair values, we decided to explore how the CV responses varied across individual subjects. Tables F.2 through F.5 present two-stage cross-sectional analyses of the four injuries associated with the use of either cleaning agent or drain opener. Each logit (maximum likelihood) equation explains whether or not a subject gave a nonzero response to the CV questions, while the OLS equations explain the variation in CV injury valuations.

We expected injury avoidance to be a normal good—that is, the CV values would be positively related to a subject's family income level. Both of the two income coefficients in the logit equations for the two drain opener injuries are positive and statistically significant (at a 90 percent confidence level), but the four income coefficients in the corresponding OLS equations are insignificant. None of the income coefficients in the four cleaning agent injury equations is significant. Given our relatively small sample size, we have insufficient evidence to support the hypothesis of a positive income effect for injury valuations derived from the CV methodology.

The *Relative risk* variable measures a subject's appraisal of his or her household's exposure relative to the average. We expected

Table F.2 CV equations for value of avoiding bleach gassings

Variable	Logit (whether value > 0)[a]	OLS (for subjects with value > 0)[b]
INCOME	-6.42×10^{-6}	-1.10×10^{-5}
	(0.20)	(−0.041)
MARRY	0.305	−3.233
	(0.44)	(−0.375)
BLACK	−0.749	−7.955
	(1.82)	(−0.765)
MALE	−0.37	3.179
	(0.63)	(0.368)
EDUC	0.230	−0.799
	(5.05)	(−0.468)
AGE	−0.006	−0.298
	(0.11)	(0.849)
Relative risk	0.018	−2.261
	(0.00)	(−0.333)
Intercept	−3.688	34.931
	(6.45)	(1.373)
R^2	—	0.06
Adj. R^2		−0.07
−2 log likelihood	153.30	—

a. Chi-square statistics are given in parentheses.
b. t-statistics are given in parentheses.

that subjects perceiving their households to be relatively risky would offer higher injury valuations than less risky households. The results in Tables F.2 through F.5 provide no support for this hypothesis. Again, this finding may be a result of the relatively small sample size.

Neither marital status, race, sex, education, nor age demonstrates strong cross-sectional effects. Two of the eight coefficients were positive and significant for the variables *MARRY, MALE,* and *EDUC*, with only one out of the eight coefficients negative and significant for *AGE*.

Table F.3 CV equations for value of avoiding bleach poisonings

Variable	Logit (whether value > 0)[a]	OLS (for subjects with value > 0)[b]
INCOME	5.9×10^{-7}	2.338×10^{-4}
	(0.00)	(−0.861)
MARRY	0.086	4.920
	(0.04)	(1.704)
BLACK	−0.664	−3.412
	(1.64)	(−0.336)
MALE	−0.629	3.313
	(1.93)	(0.318)
EDUC	0.275	1.255
	(7.56)	(0.750)
AGE	−0.013	−0.627
	(0.46)	(−1.782)
Relative risk	0.289	3.167
	(0.62)	(0.429)
Intercept	−4.001	10.583
	(8.84)	(0.439)
R^2	—	0.11
Adj. R^2	—	−0.11
−2 log likelihood	164.31	—

a. Chi-square statistics are given in parentheses.
b. t-statistics are given in parentheses.

Table F.4 CV equations for value of avoiding drain opener hand burns

Variable	Logit (whether value > 0)[a]	OLS (for subjects with value > 0)[b]
INCOME	3.67×10^{-5}	-9.64×10^{-7}
	(6.55)	(−0.056)
MARRY	0.382	1.772
	(0.50)	(2.568)
BLACK	−0.726	−0.536
	(1.63)	(0.724)
MALE	0.616	1.211
	(1.45)	(1.752)
EDUC	−0.159	−0.162
	(2.76)	(−1.288)
AGE	−0.012	−0.0321
	(0.32)	(−1.167)
Relative risk	—	−0.149
		(−0.265)
Intercept	0.490	3.789
	(0.11)	(1.742)
R^2	—	0.18
Adj. R^2	—	0.08
−2 log likelihood	135.80	—

a. Chi-square statistics are given in parentheses.
b. t-statistics are given in parentheses.

Table F.5 CV equations for value of avoiding drain opener poisonings

Variable	Logit (whether value > 0)[a]	OLS (for subjects with value > 0)[b]
INCOME	1.966×10^{-5}	-2.503×10^{-5}
	(2.58)	(−0.460)
MARRY	0.015	2.184
	(0.00)	(1.308)
BLACK	−0.104	−2.369
	(0.02)	(−1.005)
MALE	0.058	5.506
	(0.02)	(2.474)
EDUC	−0.069	0.090
	(0.69)	(0.221)
AGE	−0.015	−0.097
	(0.59)	(−1.112)
Relative risk	0.104	−0.524
	(0.07)	(−0.306)
Intercept	0.435	4.930
	(0.10)	(0.697)
R^2	—	0.15
Adj. $-R^2$	—	0.05
−2 log likelihood	152.36	—

a. Chi-square statistics are given in parentheses.
b. t-statistics are given in parentheses.

APPENDIX G

Chemical Labeling Questionnaire for Workplace Risks

LABEL STUDY QUESTIONNAIRE

Differences in background often affect the way people see
the work situation and how they feel about it. The
following questions are asked so that these differences can
be studied. The questions are not asked to identify you;
they are, in fact, designed to preserve your anonymity. If
you have any questions about your anonymity being preserved
--please do not answer them.

Now, I'd like to get some background information about you.
----

1. Are you married, widowed, separated, divorced, or have
    you never been married?

    _____ Married _____ Widowed _____ Separated
    _____ Divorced _____ Never Married

2. How many dependents do you have, excluding yourself?___

3. Sex:

    _____ Male
    _____ Female

4. Age:

    _____ 19 or under
    _____ 20-24
    _____ 25-29
    _____ 30-34
    _____ 35-39
    _____ 40-44
    _____ 45-49
    _____ 50 or over

5. Education:

    _____ Grade School
    _____ High School diploma
    _____ Some College work
    _____ Technical or associate degree
    _____ Bachelor's degree
    _____ Graduate work, no degree
    _____ Master's degree
    _____ Doctorate degre

6. Major field of study:

        _____ Biological sciences
        _____ Business or economics
        _____ Engineering
        _____ Liberal arts or humanities
        _____ Chemical or physical sciences
        _____ Technical training
        _____ Other _____

7. Now, let's talk about your present job. What is your
    general job title (e.g., supervisor, manager, engineer,
    technician, mechanic, etc.): _____

8. How many people report to your supervisor?

        _____ 1-4
        _____ 5-7
        _____ 8-10
        _____ 11-15
        _____ 16-24
        _____ 25 or over

9. How many levels of management are there between you and
    the manager of this facility?

        _____ None
        _____ One
        _____ Two
        _____ Three
        _____ Four
        _____ Over Four

10. For how many years or months, total, have you worked
    for your present employer?

        _____ Years   or   _____ Months

11. For about how long have you had the job you have now
    (working for this same employer)?

        _____ Years   or   _____ Months

12. About how many years in total have you worked for pay
    since you were 16 years old?

        _____ Number of years

13. (a) Do you supervise anyone as part of your job?

        _____ Yes   _____ No

    (b) If Yes, how many? _____

14. On this job, are you salaried or paid by the hour?

    _____ Salaried   _____ Hourly

15. If Salaried: $_____ per_____.

    If hourly: $_____ per hour.

16. Does your job at any time expose you to what you feel
    are physical dangers or unhealthy conditions?

    _____Yes   _____ No

    If Yes, then go to Question 17. If No, then go to
    Question 25.

17. What are those dangers or unhealthy conditions?

    List below the most important.

18. Dangerous or unhealthy condition #1 _____
    _____

19. How severe a problem for you is this?

    _____ No problem at all   _____ Slight

    _____ Sizable   _____ Great

20. Dangerous or unhealthy condition #2 _____
    _____

21. How severe a problem for you is this?

    _____ No problem at all   _____ Slight

    _____ Sizable   _____ Great

22. Dangerous or unhealthy condition #3 _____
    _____

23. How severe a problem for you is this?

    _____ No problem at all   _____ Slight

    _____ Sizable   _____ Great

24. When you took your job, did you feel that because of
    the nature of the Chemical Industry, the pay was
    higher than the pay in other industries?

    _____ Yes   _____ No

25. One way to measure how dangerous a job is, is to use
    the Job Danger Scale shown below. At the low end of
    the scale is the safest job, and at the high end is the
    most dangerous job.

    In order to help you judge where your job would fall on
    the scale, the ▲ on the Job Danger Scale below shows
    where the average worker in the United States feels his
    job belongs, in terms of danger.

    Please indicate where you feel your own job belongs by
    placing an X on the line between Very Safe Job and Very
    Dangerous Job.

    The more you move to the right end of the scale, the
    more dangerous you feel your job is. Just decide how
    safe or dangerous you feel your job is and place your X
    on the line as far to the right end as you feel
    reflects your feelings.

    For example, if you feel your job is totally safe,
    place your X right on the dot at the left end of the
    scale. If you feel it is pretty dangerous, place the X
    somewhere near the dot on the right end of the scale.
    (Demonstrate by pointing with a pencil as you ask the
    question.)

    ●--●
    Very   ▲   Very
    Safe   Dangerous
    Job   Job

26. Knowing what you know now, if you had to decide all
    over again whether to take the job you now have, what
    would you decide? Would you decide without any
    hesitation to take the same job, would you have some
    second thoughts, or would you decide definitely not to
    take the same job?

        _____Decide without hesitation to take same job.
        _____Have some second thoughts.
        _____Decide definitely not to take the job.

27. Taking everything into consideration, especially the
    health and safety aspects of your job, how likely is it
    that you will make a genuine effort to find a new job
    with another employer within the next year--very
    likely, somewhat likely, or not at all likely?

        _____Very likely   _____Somewhat likely

              _____ Not at all likely

LABEL PRESENTATION

Pretests have shown that each of the labels employed in this study requires a maximum of 50 seconds reading time. Therefore, to insure adequate treatment, provide Respondent (R) with label and after 60 seconds, ask the following questions:

1. What is the name of this product?

2. What do you feel is most dangerous about this product?

3. What safety precautions are required if you must work with this product?

4. Who manufactures this product?

5. What is the name of the main ingredient in this product?

If any answer is wrong or incomplete, ask R to read the label for an additional 30 seconds. Ask Questions 1, 2, 3, 4, and 5 again. Repeat until all answers are correct. Then proceed to Question 27.

Answers to Label Presentation

Correct Answers: The following answers are to be used as a guide in judging whether Respondent has read and understood the Label.

1. Label 'A' - Asbestop.

2. Causes cancer.

3. Wear mask and protective clothing, wash after handling, prevent all body contact.

4. Cuny Chemical.

5. Asbestos.

1. Label 'B' - Cunylac.

2. Irritant.

3. Wear Respirator.

4. Cuny Chemical.

5. Chloroacetophenone.

1. <u>Label 'C' CunyTnt or TNT</u>.

2. Explodes.

3. Keep cool, dry.

4. Cuny Chemical.

5. TNT or Trinitrotoluene.

                    (interviewer, circle appropriate label)

28. Now that you have seen the label for (Asbestop,
    Cunylac, Cunytnt), tell me, how would you feel about
    working with this product?

    ----

29. If your plant were to begin manufacturing and
    processing this chemical, and you had to work with this
    chemical, how dangerous do you feel your job would be?

    One way to measure how dangerous a job is, is to use
    the Job Danger Scale shown below. At the low end of
    the scale is the safest job, and at the high end is the
    most dangerous job.

    In order to help you judge where your job would fall on
    the scale, the ▲ on the Job Danger Scale below shows
    where the <u>average</u> worker in the United States feels his
    job belongs, in terms of danger.

    Please indicate where you feel your own job belongs by
    placing an X on the line between <u>Very Safe Job</u> and <u>Very
    Dangerous Job</u>.

    The more you move to the right end of the scale, the
    more dangerous you feel your job is. Just decide how
    safe or dangerous you feel your job is and place your X
    on the line as far to the right end as you feel
    reflects your feelings.

    For example, if you feel your job is totally safe,
    place your X right on the dot at the left end of the
    scale. If you feel it is pretty dangerous, place the X
    somewhere near the dot on the right end of the scale.
    (Demonstrate by pointing with a pencil as you ask the
    question.)

    ●------------------------▲------------------------------●
    Very   Very
    Safe   Dangerous
    Job   Job

30. If you had to work with this chemical on a regular basis as part of your job, would you expect to get an increase in pay?

    _____ Yes   _____ No

    If yes, how much?   Hourly: $_____ per hour
                         Salaried: $_____ per month
                              or: _____ % increase

31. Without additional compensation, if you had to work with this chemical as above, how likely is it that you would quit?

    ____ Very Likely   _____ Likely   _____ Unlikely

32. Knowing what you now know about this chemical, and if you had to work with this chemical, would you take a job that required you to work with this chemical?

    _____ Decide without hesitation to take same job.
    _____ Have some second thoughts.
    _____ Decide definitely not to take the job.

LABEL STUDY QUESTIONNAIRE - INTERVIEWER OBSERVATION

34. Respondent's (R) race: _____White   _____Black
                               _____Other   Specify:__

35. R's weight: _____Obese   _____Overweight
    _____Average for Height _____Underweight _____Skinny

36. About how tall is R?   _____Feet   _____Inches

37. How cooperative was R?   _____Very cooperative
         _____Somewhat cooperative   _____ Not cooperative

38. How well did R understand the questions?

    _____Good Understanding   _____Fair Understanding
         _____Poor Understanding

39. Did R have any speech defects or other difficulty in
    speaking English?
         _____No   _____Yes   If yes, Specify_____
                                      ----

40. Rate R's apparent intelligence.

    _____Very High _____Above Average   _____Average
         _____Below Average   _____Very Low

41. Does R have any obvious disfigurements, missing limbs,
    or habits that could make it difficult for (him/her) to
    get a job?

         _____No   _____Yes If yes, Specify_____
                             ----

42. How suspicious did R seem about the study before the
    interview?

    _____Not at all _____Somewhat _____ Very Suspicious

43. Overall, how great was R's interest in the interview?

         _____Very high _____Above Average   _____Average

         _____Below average   _____Very low

NOTES

1. Information Processing and Individual Decisions

1. For a review of this evidence, see Fischhoff and Beyth-Marom (1983), Tversky and Kahneman (1974, 1985), and Kunreuther et al. (1978). Some models of irrational behavior actually address situations that can be explained by using the correct model of rational actions. See Blomquist's (1985) analysis of precautions by auto drivers.

2. Such irrationality with respect to gambles is not unprecedented. Tversky and Kahneman (1981), for example, incorporated the overestimation of the risks of low-probability events, such as that found by Slovic, Fischhoff, and Lichtenstein (1978), into their prospect theory model of decision making.

2. Cognitive Considerations in Presenting Risk Information

1. The foundations of this approach to understanding human behavior are reviewed in Haugeland (1981) and Newell and Simon (1972), and its application to consumer choices is examined by Bettman (1979).

2. For a review of the principles of heuristic strategies see Haugeland (1981) and Card, Moran, and Newell (1983).

3. See Wilkie (1975). For an example and some data demonstrating the need for such a system for over-the-counter antacids, see Wright (1979).

3. The Design of the Consumer Information Study

1. Although we present these decisions as if they were taken sequentially, they are sometimes made simultaneously. Certainly, the product prices also affect these decisions.

2. Since we did not observe actual use behavior, usage intentions were employed as a measure of the eventual behavioral responses. Chapter 4 discusses the reasonableness of this assumption.

3. Because of the restricted size of our sample and the large number of variations already built into our design, we did not want to partition our sample further into any smaller groups. Thus in this study we were unable to examine whether placing the child poisoning questions after the chloramine gas and hand burn questions biased the subjects' responses upward or downward.

4. The Effect of Risk Information on Precautionary Behavior

1. For a pessimistic view of consumer information policies, see Adler and Pittle (1984). Related papers include Mazis, Staelin, Beales, and Salop (1981) and Staelin (1978).

2. The seminal seatbelt model was developed by Peltzman (1975). Also see Crandall and Graham (1984) and Blomquist (forthcoming). The model developed below closely resembles the safety cap model in Viscusi (1984).

3. For answering questions about which information formats induce the most precaution-taking, we need only assume that the differences in intended behavior in the survey responses translate into similar differences in the actual behavior. Lichtenstein and Slovic (1973) analyze the close similarities between hypothetical and actual decisions in gambling behavior.

4. Only the Test and Bright fractions are statistically different from the No Warning fraction at a 95 percent confidence level.

5. The differences in all three bleach precautions coefficients are significant at the 5 percent level, with the drain opener childproofing coefficients significantly different at the 10 percent level.

5. Risk-Dollar Tradeoffs, Risk Perceptions, and Consumer Behavior

1. For a discussion of this literature, see Viscusi (1986).

2. In this regression we further restrict the sample to indifferent pairs for which the risk of the new product was nonzero (that is, less than a 100 percent reduction in the number of injuries), but lower than the risk of the current product (that is, eliminating the one question in the set of paired comparisons that presents a new product more risky than the current one). In Appendix F we analyze the deleted pairs to determine if consumers are willing to pay a certainty premium to eliminate all risks from the products and if they respond differently to avoiding risk increases than to achieving risk decreases. Appendix F also explores how the valuation responses vary cross-sectionally.

3. These results provide another demonstration that the method of eliciting preferences can influence the values that consumers say they attach to a commodity or product attribute. See Hershey, Kunreuther, and Schoemaker (1982) for another example of what they call a "response mode bias."

4. The results may also appear to give somewhat implausible relative valuations of the health outcomes. For example, the CV mean value of a child poisoning from drain opener is below the value of a child poisoning from bleach, yet the drain opener poisoning is much more severe. We suspect that this seeming anomaly can be explained by order effects. Our survey always asked the child poisoning question as the second injury to be valued. Because subjects had already answered a CV question and six paired comparison questions about another injury, their answers to the child poisoning CV questions were less precise (that is, of higher variance) and could have been influenced by earlier answers. This possibility makes the comparison of the two child poisoning valuations difficult to evaluate.

6. Hazard Warnings for Workplace Risks

1. *Federal Register* (Nov. 28, 1983), 43280.
2. Viscusi (1979) also linked compensating differentials to objective risk measures, yielding comparable wage premiums. Other studies in the extensive compensating differentials literature include Thaler and Rosen (1976), Smith (1976), and Brown (1980). Smith (1979) and Viscusi (1986) provide critical surveys of this literature.
3. Educational differences and related differences in ability to perceive risks may also play a role.
4. The job hazard-quit results in Viscusi (1979) are presented for aggregative quit rates, three national samples of panel quit data (University of Michigan Panel Study of Income Dynamics and two National Longitudinal Surveys), and quit intention data from the University of Michigan's (1975) Survey of Working Conditions.
5. See Pratt, Raiffa, and Schlaifer (1975) for more detailed advocacy of the use of beta distributions for Bernoulli processes.
6. If workers do not in fact treat the label as equivalent to additional job experiences but rather "forget" their earlier knowledge, no difficulties are caused provided that the degree of forgetting is determined by the precision of their judgments, not by the level of the risk. If the initial risk level were also to affect the weight placed on the label, the empirical estimates would be biased.
7. More specifically, let V_o be the original accident severity and V_i the severity of the postwarning accident. Suppose the components of *RISK1* represent a weighted average of these risks and take the form: $RISK1 = [\gamma p V_o/(\gamma + \xi_i)] + [\xi_i s_i V_i/(\gamma + \xi_i)]$. If we set V_o equal to 1 (no loss of generality), the values of α_i and β_i are given by: $\alpha_i = \xi_i s_i V_i/(\gamma + \xi_i)$ and $\beta_i = \gamma/(\gamma + \xi_i)$. The severity-weighted results in the text follow directly.
8. Use of the fixed effects model in compensating differentials studies is not entirely new. See Brown (1980) and, more generally, Chamberlain (1982).
9. Estimation of the *EARNG* and *LNEARNG* equations for the BC/TECH subsample yielded annual risk premiums about $1,000 less than for the full sample.
10. Using the procedure developed by Chamberlain (1980), we restrict the sample to individuals who altered their quit decision, since sample observations involving the same quit responses provide no useful information for the estimation. Those (0,1) responses who would quit after the warning but not before (primarily from LAC, TNT, and ASB groups) constitute one of the binary outcomes and the (1,0) responses (primary from CARB) constitute the other outcome. The explanatory variables are the first differences of the variables included in the preinformation equation so that only the risk-related variables remain.

REFERENCES

Adler, R., and D. Pittle. 1984. Cajolery or command: are education campaigns an adequate substitute for regulation? *Yale Journal of Regulation* 2:159–194.

Arrow, K. 1974. Limited knowledge and economic analysis. *American Economic Review* 64:1–10.

———. 1982. Risk perception in psychology and economics. *Economic Inquiry* 20:1–9.

Bettman, J. R. 1979. *An Information Processing Theory of Consumer Choice.* Reading, Mass.: Addison-Wesley.

Bettman, J. R., and P. Kakkar. 1977. Effects of information presentation format on consumer information acquisition strategies. *Journal of Consumer Research* 3:233–240.

Bishop, R. C., and T. A. Heberlein. 1979. Measuring values of extramarket goods: are indirect measures biased? *American Journal of Agricultural Economics* 61:926–930.

Blomquist, G. 1985. A utility maximization model of driver traffic safety behavior. Working Paper, Department of Economics, University of Kentucky.

———. Forthcoming. *Traffic Safety Regulation by NHTSA.* Washington, D.C.: American Enterprise Institute.

Brown, C. 1980. Equalizing differences in the labor market. *Quarterly Journal of Economics* 94:113–134.

Burton, I., R. W. Yates, and G. F. White. 1978. *The Environment as a Hazard.* New York: Oxford University Press.

Card, S. K., T. P. Moran, and A. Newell. 1983. *The Psychology of Human-Computer Interaction.* Hillsdale, N.J.: Lawrence Erlbaum Associates.

Chamberlain, G. 1980. Analysis of covariance with qualitative data. *Review of Economic Studies* 47:225–238.

———. 1982. Panel data. NBER Working Paper No. 913. Cambridge, Mass.: National Bureau of Economic Research.

Chase, W. G., and K. A. Ericsson. 1981. Skilled memory. In *Cognitive Skills and Their Acquisition*, ed. J. R. Anderson. Hillsdale, N.J.: Lawrence Erlbaum Associates.

Combs, B., and P. Slovic. 1979. Causes of death: biased newspaper coverage and biased judgments. *Journalism Quarterly* 56:837–843.

Crandall, R., and J. Graham. 1984. Automobile safety regulation and offsetting behavior: some new empirical estimates. *American Economic Review* 74:328–331.

Cummings, R. G., D. S. Brookshire, and W. D. Schulze. 1984. Valuing environmental goods: a state of the art assessment of the contingent valuation method. Draft report prepared for U.S. Environmental Protection Agency, Washington, D.C.

Desvousges, W. H., V. K. Smith, and M. McGivney. 1983. A comparison of alternative approaches for estimating recreation and related benefits of water quality. Environmental Benefits Analysis Series, U.S. Environmental Protection Agency, Washington, D.C.

Fischhoff, B., and R. Beyth-Marom. 1983. Hypothesis evaluation from a Bayesian perspective. *Psychological Review* 90:239–260.

Fischhoff, B., P. Slovic, and S. Lichtenstein. 1980. Knowing what you want: measuring labile values. In *Cognitive Processes in Choice and Decision Behavior*, ed. T. S. Wallsten. Hillsdale, N.J.: Lawrence Erlbaum Associates.

Green, P. E., and V. Srinivasan. 1978. Conjoint analysis in consumer research: issues and outlook. *Journal of Consumer Research* 5:103–123.

Hadden, S. G. 1985. *Read the Label: Reducing Risk by Providing Information*. St. Paul, Minn.: Westview Publishing Co.

Haugeland, J., ed. 1981. *Mind Design*. Cambridge, Mass.: MIT Press.

Hershey, J. C., H. C. Kunreuther, and P. J. H. Schoemaker. 1982. Sources of bias in assessment procedures for utility functions. *Management Science* 28:936–954.

Johnson, E. J., and J. W. Payne. 1985. Effort and accuracy in choice. *Management Science* 31:395–414.

Kahneman, D., and A. Tversky. 1979. Prospect theory: an analysis of decision under risk. *Econometrica* 47:263–291.

Kahneman, D., P. Slovic, and A. Tversky, eds. 1982. *Judgment under Uncertainty: Heuristics and Biases*. Cambridge: Cambridge University Press.

Kanouse, D. E., and B. Hayes-Roth. 1980. Cognitive considerations in the design of product warnings. In *Product Labeling and Health Risks*, ed. L. A. Morris, M. Mazis, and I. Barofsky. Cold Spring Harbor, N.Y.: Cold Spring Harbor Laboratory.

Kates, R. W. 1982. Hazard and choice perception in flood plain management. Research Paper No. 78, Department of Geography, University of Chicago.

Keller, K. L., and R. Staelin. 1985. Effects of quality and quantity of information on decision effectiveness. Working Paper, Fuqua School of Business, Duke University.

Kunreuther, H. 1976. Limited knowledge and insurance protection. *Public Policy* 24:227–261.

Kunreuther, H., R. Ginsberg, L. Miller, P. Sagi, P. Slovic, B. Borkin, and N. Katz. 1978. *Disaster Insurance Protection: Public Policy Lessons*. New York: Wiley.

Ley, P. 1980. Practical methods of improving communication. In *Product Labeling and Health Risks*, ed. L. A. Morris, M. Mazis, and I. Barofsky. Cold Spring Harbor, N.Y.: Cold Spring Harbor Laboratory.

Lichtenstein, S., and P. Slovic. 1973. Response-induced reversals of preferences in gambling: an extended replication in Las Vegas. *Journal of Experimental Psychology* 101:16–20.

Lichtenstein, S., P. Slovic, B. Fischhoff, M. Layman, and B. Combs. 1978.

Judged frequency of lethal events. *Journal of Experimental Psychology: Human Learning and Memory* 4:551–78.

Magat, W. A., J. W. Payne, and P. F. Brucato. 1985. Information provision as a regulatory alternative: lessons from home energy audit programs. Working Paper #84-3, Duke University Center for the Study of Business Regulation.

Magat, W. A., W. K. Viscusi, and J. Huber. 1985. Comparing alternative non-market approaches to morbidity risk valuation. Working Paper #85-4, Duke University Center for the Study of Business Regulation.

Malhotra, N. K. 1982. Information load and consumer decision making. *Journal of Consumer Research* 8:419–430.

Mazis, M. B., R. Staelin, H. Beales, and S. Salop. 1981. A framework for evaluating consumer information regulation. *Journal of Marketing* 45:11–21.

Michigan, University of. 1975. Institute of Social Research *Survey of Working Conditions*, SRC Study No. 45369. Ann Arbor, Mich.: University of Michigan Social Science Archive.

Miller, G. A. 1956. The magical number seven, plus or minus two: some limits on our capacity for processing information. *Psychological Review* 63:81–97.

Muller, T. E. 1982. The impact of consumer information on brand sales: a field experiment with point of purchase nutritional information load. Unpublished doctoral dissertation, University of British Columbia.

Newell, A., and H. A. Simon. 1972. *Human Problem Solving.* Englewood Cliffs, N.J.: Prentice-Hall.

———. 1981. Computer science as empirical inquiry: symbols and search. In *Mind Design*, ed. J. Haugeland. Cambridge, Mass.: MIT Press.

O'Connor, C., and S. Lirtzman. 1984. *Handbook of Chemical Industry Labeling.* Park Ridge, N.J.: Noyes Publications.

Payne, J. W. 1976. Task complexity and contingent processing in decision making: an information search and protocol analysis. *Organizational Behavior and Human Performance* 16:366–387.

———. 1982. Contingent decision behavior. *Psychological Bulletin* 92:382–402.

———. 1985. The psychology of risk taking. In *Behavioral Decision Making*, ed. G. Wright. New York: Plenum.

Peltzman, S. 1975. The effects of automobile safety regulation. *Journal of Political Economy* 83:677–725.

Pratt, J., H. Raiffa, and R. Schlaifer. 1975. *Introduction to Statistical Decision Theory.* New York: McGraw-Hill.

Raiffa, H. 1968. *Decision Analysis.* Reading, Mass.: Addison-Wesley.

Randall, A., J. P. Hoehn, and D. S. Brookshire. 1983. Contingent valuation surveys for valuing environmental assets. *Natural Resources Journal* 23:637–648.

Reed, S. K. 1982. *Cognition: Theory and Application.* Monterey, Calif.: Brooks/Cole.

Rethans, A. 1979. An investigation of consumer perceptions of product hazards. Doctoral dissertation, University of Oregon-Eugene.

Russo, J. E. 1974. More information is better: a reevaluation of Jacoby, Speller, and Kohn. *Journal of Consumer Research* 1:68–72.

———. 1977. The value of unit price information. *Journal of Marketing Research* 14:193–201.

Russo, J. E., G. Krieser, and S. Miyashita. 1975. An effective display of unit price information. *Journal of Marketing* 39:11–19.

Russo, J. E., R. Staelin, C. Nolan, G. Russell, and B. Metcalf. 1986. Nutrition information in the supermarket. *Journal of Consumer Behavior* 13:48–70.

Simon, H. A. 1974. How big is a chunk? *Science* 183:482–488.

———. 1981. *The Sciences of the Artificial*, 2nd ed. Cambridge, Mass.: MIT Press.

Slovic, P., B. Fischhoff, and S. Lichtenstein. 1978. Accident probabilities and seat belt usage: a psychological perspective. *Accident Analysis and Prevention* 10:281.

———. 1980. Informing people about risk. In *Product Labeling and Health Risks*, ed. L. A. Morris, M. Mazis, and I. Barofsky. Cold Spring Harbor, N.Y.: Cold Spring Harbor Laboratory.

———. 1982. Facts versus fears: understanding perceived risk. In *Judgment under Uncertainty: Heuristics and Biases*, ed. D. Kahneman, P. Slovic, and A. Tversky. Cambridge: Cambridge University Press.

Smith, R. S. 1976. *The Occupational Safety and Health Act: Its Goals and Achievements*. Washington, D.C.: American Enterprise Institute.

———. 1979. Compensating differentials and public policy: a review. *Industrial and Labor Relations Review* 32:339–352.

Smith, V., W. Desvousges, and A. Freeman III. 1985. Valuing changes in hazardous waste risks: a contingent valuation analysis. Benefits Branch, Economic Analysis Division, U.S. Environmental Protection Agency, Washington, D.C.

Staelin, R. 1978. The effects of consumer education on consumer product safety behavior. *Journal of Consumer Research* 5:30–40.

Svenson, O. 1979. Are we all among the better drivers? Unpublished manuscript, Department of Psychology, University of Stockholm.

Thaler, R., and S. Rosen. 1976. The value of saving a life: evidence from the labor market. In *Household Production and Consumption*, ed. N. Terleckyj. NBER Studies in Income and Wealth No. 40, New York: Columbia University Press.

Tversky, A., and D. Kahneman. 1981. The framing of decisions and the psychology of choice. *Science* 211:453–458.

———. 1982. Judgment under uncertainty: heuristics and biases. In *Judgment under Uncertainty: Heuristics and Biases*, ed. D. Kahneman, P. Slovic, and A. Tversky. Cambridge: Cambridge University Press.

———. 1985. Rational choice and the framing of decisions. Presented at the Conference on the Behavioral Foundations of Economic Theory, University of Chicago.

University of Michigan, Institute for Social Research. 1975. *Survey of Working Conditions, 1969–70*. Ann Arbor: Institute for Social Research.

Viscusi, W. K. 1979. *Employment Hazards: An Investigation of Market Performance*. Cambridge, Mass.: Harvard University Press.

———. 1983. *Risk by Choice: Regulating Health Safety in the Workplace*. Cambridge: Harvard University Press.

———. 1984. The lulling effect: the impact of child-resistant packaging on aspirin and analgesic ingestions. *American Economic Review Papers and Proceedings* 74:324–327.

———. 1985. A Bayesian perspective on biases in risk perception. *Economics Letters* 17:59–62.

———. 1986. The valuation of risks to life and health: guidelines for policy analysis. In *Proceedings of NSF Conference on Benefit Assessment: State of the Art*. Dordrecht, Holland: Reidel Publishers.

Viscusi, W. K., and W. A. Magat. 1985. Analysis of economic benefits of improved information. Draft report prepared for U.S. Environmental Protection Agency, Washington, D.C.

———. 1986. Analysis of economic benefits of improved information: project period two report. Draft report prepared for U.S. Environmental Protection Agency, Washington, D.C.

Viscusi, W. K., and C. O'Connor. 1984. Adaptive responses to chemical labeling: are workers Bayesian decision makers? *American Economic Review* 74:942–956.

Wilkie, W. L. 1975. *How Consumers Use Product Information: An Assessment of Research in Relation to Public Policy Needs*. Washington, D.C.: United States Government Printing Office.

Winett, R. A., and J. H. Kagel. 1984. Effects of information presentation format on resource use in field settings. *Journal of Consumer Research* 11:655–667.

Wright, P. L. 1979. Concrete action plans in TV messages to increase reading of drug warnings. *Journal of Consumer Research* 6:256–269.

Young, F. W. 1981. Quantitative analysis of qualitative data. *Psychometrica* 46:357–388.

INDEX